EXS 88

Proteomics in Functional Genomics

Protein Structure Analysis

Edited by P. Jollès and H. Jörnvall

Birkhäuser Verlag
Basel · Boston · Berlin

Editors

Prof. Dr. P. Jollès
Laboratoire de Chimie
des Substances Naturelles
URA C.N.R.S. No. 401
Muséum National d'Histoire Naturelle
63, rue Buffon
F-75005 Paris
France

Prof. Dr. H. Jörnvall
Department of Medical
Biochemistry and Biophysics
Karolinska Institutet
SE-171 77 Stockholm
Sweden

Library of Congress Cataloging-in-Publication Data
Proteomics in functional genomics : protein structure analysis / edited by P. Jollès, H. Jörnvall.
   p. ; cm. – (EXS ; 88)
  Includes bibliographical references and index.
  ISBN 3764358858 (hard cover : alk. paper)
  1. Proteins – Structure. 2. Proteins – Analysis. I. Title: Protein structure analysis. II.
Jollès, Pierre, 1927-III. Jörnvall, Hans. IV. Series.
  [DNLM: 1. Proteins – chemistry. 2. Genome. 3. Protein Structure. 4. Sequence Analysis.
QU 55 P9825 2000]
QP551.P7567 2000
572'.633-dc21

99-059890

Deutsche Bibliothek Cataloging-in-Publication Data
Proteomics in functional genomics : protein structure analysis / ed. by P. Jollès ; H. Jörnvall. –
Basel ; Boston ; Berlin : Birkhäuser, 2000
  (EXS ; 88)
  ISBN 3-7643-5885-8

ISBN 3-7643-5885-8 Birkhäuser Verlag, Basel – Boston – Berlin

Contents

List of Contributors

Ettore Appella, Laboratory of Cell Biology, Division of Basic Sciences, National Cancer Institute, National Institutes of Health, 37 Convent Drive, Bethesda, MD 20892, USA; e-mail: appellae@dino.nci.nih.gov

David Arnott, Protein Chemistry Department, Genentech, Inc., 1 DNA Way, South San Francisco, CA 94080, USA; e-mail: arnott@gene.com

Tomas Bergman, Department of Medical Biochemistry and Biophysics, Karolinska Institutet, SE-171 77 Stockholm, Sweden; e-mail: Tomas.Bergman@mbb.ki.se

Victoria L. Boyd, CELERA, 850 Lincoln Centre Drive, Foster City, CA 94404, USA; e-mail: boydvn@fc.celera.com

MeriLisa Bozzini, PE Biosystems, 400 Lincoln Centre Drive, Foster City, CA 94404, USA; e-mail: bozzinml@pebio.com

Julio E. Celis, Department of Medical Biochemistry and Danish Centre for Human Genome Research, Aarhus University, Ole Worms Alle, Build. 170, DK-8000 Aarhus C, Denmark; e-mail: jec@biokemi.au.dk

David R. Dupont, PE Biosystems, 700 Lincoln Centre Drive, Foster City, CA 94404, USA; e-mail: dupontdr@pebio.com

Gerrit J. Gerwig, Bijvoet Center for Biomolecular Research, Department of Bio-Organic Chemistry, Utrecht University, P.O. Box 80075, NL-3508 TB Utrecht, The Netherlands; e-mail: gerwig@boc.chem.uu.nl

Kris Gevaert, Flanders Interuniversity Institute for Biotechnology, VIB09, Department of Biochemistry, Universiteit Ghent, B-9000 Ghent, Belgium; e-mail: Kris.Gevaert@rug.ac.be

William J. Griffiths, Department of Medical Biochemistry and Biophysics, Karolinska Institutet, SE-171 77 Stockholm, Sweden; e-mail: william.griffiths@mbb.ki.se

Pavel Gromov, Department of Medical Biochemistry and Danish Centre for Human Genome Research, Aarhus University, Ole Worms Alle, Build. 170, DK-8000 Aarhus C, Denmark

Magnus Gustafsson, Department of Medical Biochemistry and Biophysics,
     Karolinska Institutet, SE-171 77 Stockholm, Sweden;
     e-mail: magnus.gustafsson@mbb.ki.se

Ulf Hellman, Ludwig Institute for Cancer Research, P.O. Box 595,
     SE-751 24 Uppsala, Sweden; e-mail: ulf.hellman@licr.uu.se

Tony Houthaeve, Flanders Interuniversity Institute for Biotechnology,
     VIB09, Department of Biochemistry, Universiteit Ghent,
     B-9000 Ghent, Belgium; e-mail: Tony.Houthaeve@rug.ac.be

Peter James, Protein Chemistry Laboratory, Institute for Biochemistry,
     Universitätsstrasse 16, ETH-Zentrum, CH-8092 Zürich, Switzerland;
     e-mail: peter.james@bc.biol.ethz.ch

Jan Johansson, Department of Medical Biochemistry and Biophysics,
     Karolinska Institutet, SE-171 77 Stockholm, Sweden;
     e-mail: jan.johansson@mbb.ki.se

Pierre Jollès, Laboratoire de Chimie des Substances Naturelles,
     URA C.N.R.S. No. 401, Muséum National d'Histoire Naturelle,
     63, rue Buffon, F-75005 Paris, France

Hans Jörnvall, Department of Medical Biochemistry and Biophysics,
     Karolinska Institutet, SE-171 77 Stockholm, Sweden

Bengt Persson, Stockholm Bioinformatic Centre and Department
     of Medical Biochemistry and Biophysics, Karolinska Institutet,
     SE-171 77 Stockholm, Sweden; e-mail: bengt.persson@mbb.ki.se

Manfredo Quadroni, Biomedical Research Center, University of British
     Columbia, 2222 Health Sciences Mall, Vancouver, B.C., Canada V6T 1Z3

Peter Roepstorff, Department of Molecular Biology, Odense University,
     DK-5230 Odense M, Denmark; e-mail:roe@mail.dou.dk

Kazuyasu Sakaguchi, Laboratory of Cell Biology, Division of Basic
     Sciences, National Cancer Institute, National Institutes of Health,
     37 Convent Drive, Bethesda, MD 20892, USA;
     e-mail: kazu1scc@mbox.nc.kyushu-u.ac.jp

John E. Shively, Division of Immunology, Beckman Research Institute
     of the City of Hope, Duarte, CA 91010, USA;
     e-mail: jshively@coh.org

Margareta Stark, Department of Medical Biochemistry and Biophysics,
   Karolinska Institutet, SE-171 77 Stockholm, Sweden;
   e-mail: margareta.stark@mbb.ki.se

Joël Vandekerckhove, Flanders Interuniversity Institute for Biotechnology,
   VIB09, Department of Biochemistry, Universiteit Ghent, B-9000 Ghent,
   Belgium; e-mail: Joel.Vandekerckhove@rug.ac.be

Johannes F. G. Vliegenthart, Bijvoet Center for Biomolecular Research,
   Department of Bio-Organic Chemistry, Utrecht University,
   P.O. Box 80075, NL-3508 TB Utrecht, The Netherlands;
   e-mail: vlieg@pobox.uu.nl

Yuqin Wang, Department of Medical Biochemistry and Biophysics,
   Karolinska Institutet, SE-171 77 Stockholm, Sweden;
   e-mail: yuqin.wang@mbb.ki.se

Peter J. Wirth, Laboratory of Experimental Carcinogenesis, Division of
   Basic Sciences, National Cancer Institute, National Institutes of Health,
   37 Convent Drive, Bethesda, MD 20892, USA

Brigitte Wittmann-Liebold, WITA GmbH, Wittmann Institute of Technology and Analysis of Biomolecules, Warthestr. 21,
   D-14513 Teltow, Germany; e-mail: info@wita.de

Christian Wurzel, WITA GmbH, Wittmann Institute of Technology and
   Analysis of Biomolecules, Warthestr. 21, D-14513 Teltow, Germany;
   e-mail: info@wita.de

Shahparak Zaltash, Department of Medical Biochemistry and Biophysics,
   Karolinska Institutet, SE-171 77 Stockholm, Sweden;
   e-mail: shahparak.zaltash@mbb.ki.se

Protein structure analysis of today: proteomics in functional genomics

With 20-odd genome projects completed and a similar number in the pipe-line, including the complete human genome, the structures essential to all life forms are now known.

With this wealth of structural information available, bioscience will make a large leap forward and forever change the pattern in at least two senses regarding protein studies. The first change is that the whole field of protein science will alter emphasis: instead of basic characterization of novel proteins, meaning full analysis of all regions and subsequent data-bank depositions, it will involve identification of already known forms, meaning limited analysis of certain regions for screening against existing databank entries. Although in one sense "easier", this shift is still push-ing the limits, since demands on sensitivity, accuracy and speed are con-tinuously increasing.

Similarly, truncations, modifications and other processings should still be detected, as should also the presence of isoforms and other variants. Thus, overall structures and compatibilities will continuously be evaluated. To meet all this, the analytical method of choice will often be mass spectro-metry.

The second change to protein chemistry is the need for simultaneous evaluation of many proteins in highly complex mixtures. Thus, not only single proteins need to be identified one at a time to understand the func-tional interactions, but rather the global output of gene products from essentially all tissues, and not only in normal cases, but also in diseases, development and other special states. Determination of global cellular protein output is a major constituent of the science of proteomics and an integral part of functional genomics in the now poststructural era.

Much as the first change in current protein analysis emphasizes mass determinations and overall compatibilities in identification, the second change emphasizes separation techniques and evaluation of patterns to eventually judge all protein products and their ratios and modifications. The methods then concern separation, and in particular efficient separation over wide charge and mass ratios. The method of choice at present is two-dimensional (2D) gel electrophoresis with pH gradients in one dimension and size gradients in the other. With the availability of preformed pH gradients, reproducibility has been secured, and a huge number of protein products can be separated and compared under various conditions. In this manner, 2D gel electrophoresis combined with mass spectrometric protein identification has rapidly become the leading methodology in present pro-tein analysis, or proteome research, as evidenced by the introductory parts of several of the chapters in this book. True, this technique is at the current front line of research, and allows separation and identification of proteins from tissue extracts on single plates and at high sensitivity. Essentially, all spots properly visible by silver straining can be identified – a true achievement, still time-consuming and costly, but feasible! In this manner, tumour marker proteins, developmental key proteins, genetic markers and other forms with special functions are now increasingly clarified. The methodologies behind much of this progress are treated by most chapters in this volume. The prospects are fantastic, and we are now witnessing an explosion in the knowledge of cellular patterns in protein expression as an integral part of understanding the interplays in functional genomics.

Nevertheless, to this enthusiasm over novel methodological combinations, perhaps a word of caution should be added. One concerns limits and sensitivity: although current techniques cover essentially all silver-stainable proteins and resolution of thousands of forms, and although this is great and "modern", it is just the start. Actually identifying THE regulating forms in global cellular protein outputs will probably have to involve sub-silver-staining amounts and tens of thousands of proteins separated. Hence, the present 2D gel electrophoresis/mass spectrometric identifications may constitute the start, but perhaps not the final word in identification of key proteins. As often before, column chromatographies and miniaturization may soon figure here, too. So perhaps 2D gel separations will not be the single method of choice in the future, and as outlined in several chapters, additional methodologies will emerge, updating micro-HPLC, capillary electrophoresis and additional column methods.

With these words of introduction, the outline and the background to several of the chapters in this volume have been given. In addition to the general background to the chapters, and to the possibly further chromatographic developments to come, one more aspect is pertinent: the importance of minor variabilities to cover not only "pure" protein chemistry, but also glycoprotein chemistry, lipoprotein chemistry and other complex protein chemistries. To reflect this general need for versatility in all analyses, and the importance of minor details in present-day 2D gel/mass spectrometry approaches, we have largely kept the original introductions from each chapter, rather than replace them with a summarized outline for each approach. This keeping of the separate introductions emphasizing similar analyses in each approach may at first give a "repetitive" and "unedited" impression to the reader of successive chapters. We apologize for this, but consider the advantage of this approach to outweigh the disadvantage of some seemingly repetitive presentations. Thus, minor variabilities in the otherwise similar approaches are highlighted, and the combined spread in methodological detail can be evaluated. Similarly, the current general applicability of the 2D gel/mass spectrometry approach can be directly grasped through the similar introductions/methodologies in different chapters.

We think the present trends come through in the chapters, and we are grateful to all authors for their excellent assembly of the major approaches in each subfield. Some true redundancy may remain, but is hopefully compensated by the multiple demonstration of each variable. It is interesting to see the common platform in current protein chemistry/proteomics analysis, and although we have also hinted at the need for and possibilities of further methodological breakthroughs, it is clear that 2D gel/mass spectrometry analysis is a central method and will remain so for a long time, as exemplified by the chapters in this volume.

Stockholm and Paris, early 2000   Hans Jörnvall and
                                                               Pierre Jollès

Proteomics in Functional Genomics
ed. by P. Jollès and H. Jörnvall
© 2000 Birkhäuser Verlag Basel/Switzerland

Proteome mapping by two-dimensional polyacrylamide gel electrophoresis in combination with mass spectrometric protein sequence analysis

Ettore Appella[1], David Arnott[2], Kazuyasu Sakaguchi[1] and Peter J. Wirth[3]

[1] *Laboratory of Cell Biology and* [3]*Laboratory of Experimental Carcinogenesis, Division of Basic Sciences, National Cancer Institute, National Institutes of Health, 37 Convent Drive, Bethesda, MD 20892, USA and* [2]*Protein Chemistry Department, Genentech, Inc., 1 DNA Way, South San Francisco, CA 94080, USA*

Summary. The high resolving power of two-dimensional polyacrylamide gel electrophoresis 2D-PAGE and its full analytical and preparative potential have been described with special emphasis on reproducibility and standardization of protein spot patterns, enhanced protein detection sensitivity, and computer analysis database development. New methodologies for peptide mass fingerprinting, peptide, sequence, and fragmention tagging have been highlighted. Major challenges associated with 2D-PAGE/mass spectrometric protein sequencing were outlined which need to be addressed in the future, including sample enrichment, use of alternative gel matrices, improvements in separation systems interfaced directly to the mass spectrometer, and design of high-sensitivity instruments with very high mass ranges. It is hoped that comparative studies to identify, quantitate, and characterize proteins differentially expressed in normal versus diseased cells would give insight into mechanisms of pathogenesis and allow the development of a way to control both the etiology and the course of diseases.

During the last few years considerable effort has gone into genomic studies and large-scale DNA-sequencing projects. The ultimate goal of these studies is to define the genome, in particular, the human genome, with the hope that one will be able to identify and study key genes functional in normal as well as in aberrant physiological pathways, including cancer development. However, estimates have been made that only 2% of human diseases result from single gene defects (i.e. expression of an altered gene or loss of function of a normal gene) [1]. In the remaining 98% of human diseases, epigenetic and environmental factors are involved that affect both the etiology and the severity of the affliction. For the development of complex diseases such as cancer, several genetic alterations must occur [2]. A molecular understanding of how cells operate during normal or pathological states requires a knowledge of which genes are expressed. While gene expression has traditionally been studied at both the messenger RNA (mRNA) level and the protein expression level, it should be emphasized that mRNA-based studies only measure message abundance and not actual protein levels which, in essence, are the functional molecules within the cell. In fact, Anderson and Seilhamer [3] have recently shown that in the human liver there may not be a good correlation between mRNA abundance and amount of protein present. Protein expression may provide a better assessment of the metabolic state of cells.

Protein analysis, however, is considerably more challenging than DNA/ RNA-based analyses, since, in contrast to DNA, which is composed of only four different nucleotides, proteins are composed of at least 20 unmodified, and many more modified, amino acids; thus the physicochemical characteristics of proteins vary considerably. Adding to the complexity of analysis is the fact that protein expression levels cover an extremely large range, and many proteins undergo a myriad of co- and posttranslational modifications, including phosphorylation, acetylation, sulfation, glycosylation, myristoylation, conjugation with lipids, and proteolytic processing, thereby further negating the "one-gene, one-polypeptide" paradigm.

In contrast to the genome, which is a relatively fixed characteristic of an organism, the proteome (*prot*ein complement expressed by the gen*ome* in a particular cell or tissue [4]) is in a constant state of flux and is dependent upon a multitude of factors, including the developmental state of the organism, tissue and organelle location, and metabolic state. Comparative studies to identify, quantitate, and characterize proteins differentially expressed in normal versus diseased cells will give insight into mechanisms of pathogenesis.

1. 2D-PAGE

The electrophoretic separation of proteins and polypeptides in polyacrylamide gels has become the method of choice for the fractionation and characterization of complex mixtures of proteins and polypeptides at both the analytical and preparative level. Separation is dependent upon the molecular weight (M_r) and shape, as well as the net charge or isoelectric point (pI) of the component proteins, and hence is directly related to the amino acid composition of the individual proteins. In practice, however, proteins are routinely separated as a function of either their M_r (molecular sieving), independent of their pI, or alternatively on the basis of their pI, independent of their M_r. Although the resolving power of either parameter is quite good, on the order of 100 distinct protein bands, this may be inadequate for many applications. For example, it has been estimated that the total number of proteins expressed in a typical mammalian cell (e.g. hepatocyte) is on the order of 5000–20,000 proteins, with as few as 3000–6000 playing significant roles in cellular maintenance. No single unidimensional separation procedure provides the necessary resolution for the analysis of highly complex, whole cell lysate protein mixtures. To circumvent this problem, Kenrick and Margolis [5] combined two commonly used electrophoretic procedures, namely, native isoelectric focusing (IEF) electrophoresis with pore gradient SDS/polyacrylamide gel electrophoresis (PAGE) to separate serum proteins on a two-dimensional (2D) pI-M_r matrix according to two independent physicochemical parameters, i.e. charge and size. Patrick O'Farrell [6], Joachim Klose [7], and George Scheele [8] independently modified this basic 2D procedure and introduced what today is commonly referred to as high-resolution 2D-PAGE. Their modifications utilize IEF under denaturing conditions (urea, nonionic detergent, and reducing agent) in the first dimension followed by molecular sieving in SDS/PAGE in the second dimension (90 degrees from the first) to resolve 2000–3000 individual polypeptides on a single electropherogram.

Recently, other high-resolution protein separation techniques, such as capillary zone electrophoresis (CZE) [9], serial high-performance liquid-liquid chromatography [10], or liquid chromatography-capillary electrophoresis [11] have been introduced, which offer promising alternatives to 2D-PAGE, but these are still in developmental stages, and to the best of our knowledge do not yet provide reproducible and complete separation of complex protein mixtures.

1.1 Advantages/strengths of 2D-PAGE

Since 2D-PAGE separations are based solely on the physicochemical (pI and M_r) characteristics of individual proteins, no functional assay is required to follow the expression of a particular protein(s) under study. This is especially important since the biological function(s) of the vast majority of cellular proteins is unknown. 2D-PAGE therefore allows one to survey simultaneously a very large number of cellular proteins in the absence of any prior knowledge concerning either the identity or biological function of the protein(s) under investigation. Thus, one is able to obtain a virtual "snapshot" representation of the steady-state level of protein expression in a cell or tissue preparation under specific metabolic conditions. Subtractive analysis of quantitative protein expression patterns (2D maps) exhibited by cellular/tissue lysates generated under different experimental conditions (e.g. following various drug/chemical treatments, stress induction, etc.) or pathological/disease states (cancer development) can be compared in order to detect protein alterations associated with a specific biological condition.

In addition to these essentially descriptive analytical applications, 2D-PAGE has recently been adapted micropreparatively for the one-step purification and isolation of homogeneously pure proteins directly from 2D gels for subsequent biochemical and chemical analysis. Protein fractions, eluted either directly from gel sections or following transfer to various membrane supports (e.g. Polyvinylidene difluoride (PVDF) membranes), have been used for the production of mono- and polyclonal antibodies directed against the isolated protein(s), analysis of receptor binding and enzymatic activities, and structural characterization. A schematic representation of 2D-PAGE and various ancillary procedures is illustrated in Figure 1.

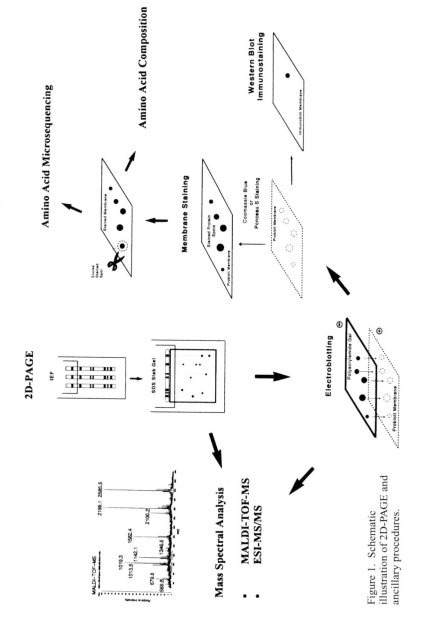

Figure 1. Schematic illustration of 2D-PAGE and ancillary procedures.

1.2 Drawbacks/limitations

Despite the high resolving power of 2D-PAGE, its full analytical and pre-parative potential has yet to be realized because of certain technical lim-itations. For instance, intrinsic drawbacks in the use of synthetic carrier ampholytes (CAs) for the generation and stabilization of the first-dimen-sion pH gradients during IEF have been troublesome. Because of the highly heterogeneous and generally poorly defined chemical composition of com-mercially available CAs, considerable variability in protein spot patterns is often encountered, especially when using CAs from different manufac-turers. This can result in serious difficulties in gel pattern matching and protein identification, thereby making interlaboratory and intralaboratory comparisons difficult. In addition, the pH gradients generated utilizing CAs are subject to severe cathodic drift at pH values > 8.0 due to electro-endosmotic effects, thus preventing the separation and analysis of the more basic proteins (e.g. DNA-binding proteins, transcriptional factors, ribo-somal proteins, and histones). A less frequently encountered but none-theless troublesome complication involves the reactivity/and or interaction of individual proteins with the CAs themselves, resulting in artifactual protein-CA spots.

While not exclusively limited to 2D-PAGE *per se*, the methodology also suffers from the relatively insensitive procedures available for protein spot detection in general. Visualization of individual proteins from 2D-PAGE gels is usually performed postelectrophoretically using organic dyes and heavy metal stains, such as Coomassie Blue with silver "shadowing" [12]. In contrast to genetic studies in which powerful enrichment procedures [i.e. polymerase chain reaction (PCR)] exist for the amplification of minute quantities of DNA and RNA, no analogous procedures exist for the am-plification of proteins. The lower limit for spot detection and visualization using ultrasensitive silver staining is on the order of 1 ng (~ 0.1 pmol for 10 kDa protein). Therefore, only proteins expressed at levels greater than 10^3 copies per cell can be detected [13]. This relatively "low" detection sen-sitivity has necessitated the utilization of additional time-consuming pro-tein enrichment strategies, including subcellular fractionaction, multiple 2D gels, and the pooling of multiple protein spots, to obtain sufficient quantities for biochemical characterization.

2. Technological improvements

Numerous technological improvements and modifications have been made to the original 2D-PAGE protocol. Approaches have focused on the devel-opment of reproducible and standardized procedures for sample prepara-tion and the running of gels, enhanced spot detection and accurate protein quantitation, development of computer software for the analysis of polypeptide spot patterns and database construction, as well as the implementation of protocols interfacing 2D-PAGE with methodology for protein structure characterization.

2.1 Reproducibility and standardization of spot patterns

Recently, immobilized pH gradient (IPG) electrophoresis has been introduced to eliminate complications associated with the use of CAs in the first dimension IEF mode of 2D-PAGE. This procedure utilizes ultrathin (0.2 mm thick) precast polyacrylamide gel strips containing covalently bound low M_r monosubstituted polyacrylamido-acids and bases to form well-defined, highly stable, preformed pH gradients [14–16]. IPG gel strips (Pharmacia), formulated in a variety of narrow (pH <1) and broad (pH 3–10) range gradients, offer many advantages not afforded by CA-IEF that include (i) enhanced gel pattern reproducibility, thereby greatly facilitating inter- and intralaboratory gel pattern comparisons; (ii) increased spot resolution and detection sensitivity over wide pH ranges; (iii) enhanced loading capacity for subsequent micropreparative applications. Whereas CA-IEF is essentially limited to the separation of microgram quantities of proteins between pH 4.5 and 7.5, IPG-2D-PAGE permits the separation of milligram (1–15 mg) quantities of proteins on a single IPG strip [17, 18]. Acidic proteins (pI < 4) as well as highly basic proteins, including those in the pI range of 10–12 [19], can routinely be separated using IPG on a single 2D gel. Such separations using either CA-IEF or nonequilibrium pH gradient electrophoresis (NEPHGE) are virtually impossible, due to heavy protein streaking in basic pH regions as a result of CA-associated cathodic drift or the highly variable nature of NEPHGE [20]. In addition, IPG strips are considerably easier to manipulate than tube gels, since the IPG gels are covalently bound to semirigid mylar backing strips and are not subject to gel stretching and breakage, thereby minimizing geometric distortions in gel patterns. Studies within a single laboratory [21] or among laboratories [22–24] comparing the reproducibility of 2D-PAGE patterns using IPG gels in the first dimension showed less than a 1-mm positional variability (x-y axis) between spots observed in standardized protein samples.

2.2 Enhanced spot detection/sensitivity

Since most physiological and pathological processes are associated with a quantitative modulation of gene expression, more specifically protein expression, sensitive spot detection and accurate protein quantitation methodologies are of utmost importance. The use of broad range (pH 4–12) IPGs provides an excellent global overview of protein composition, whereas separations utilizing well-defined, narrow range (1–2 pH unit)

IPGs, specifically designed for the protein(s) of interest, have permitted the detection and analysis of individual polypeptides and polypeptide isoforms (e.g. phosphoproteins, glycoproteins, etc.) having pIs differing by a little as 0.01 pH units.

While the lower limit of spot detection using current silver staining methods is approximately 1 ng (~0.1 pmol for a 10-kDa protein) [25] or roughly 10^3 copies per cell [13], immunoblotting, using high-affinity mono-clonal antibodies and visualization with enhanced chemiluminescence (ECL), allows detection of low-abundance proteins present at levels as low as 1–10 copies per cell (1.7 fmol). Sanchez and co-workers utilized mul-tiple 2D immunoblotting with a mixture of nine monoclonal antibodies (p53, c-myc, PCNA, MEK1, pan-ras, Cip1, Cdc2, Kip1, and TCTP) to identify and map oncogene expression and cell cycle-specific checkpoints in both patient solid biopsy samples and transformed cell lines [26]. The antibody mixture simultaneously detected and provided quantitative infor-mation on all nine proteins as well as identifying the presence of various isoforms that most likely result from posttranslational modifications of the primary gene products.

Greater spot detection sensitivity and more accurate quantitation can be achieved with the use of metabolic labeling, either pre- or postelectro-phoretically, with radioisotope [^{35}S, ^{14}C, ^{3}H, ^{32}P, or ^{125}I]-containing precur-sor molecules. The recent introduction of phosphor-imaging technology has further enhanced detection sensitivity and quantitation by greatly in-creasing the dynamic linearity of radiolabel detection (four orders of mag-nitude) and reducing exposure time from 1 month (fluorography-multiple exposures) to 3–5 days [27].

2.3 Computer analysis/database development/WWW access

Although 2D-PAGE maps appear complex and complicated, the technique is highly reproducible, and simple analysis of the 2D-PAGE patterns can be performed by superimposing one photographic image over another on a light box. For more complex studies, sophisticated computer software packages have been developed to aid in the scanning and digitalization of 2D-PAGE maps, including the segmentation and quantitation of individual protein spots and the automatic matching of gel images [28–33]. The high reproducibility of IPG-2D-PAGE has permitted the generation of several high-resolution 2D-PAGE reference maps and comprehensive databases of qualitative and quantitative protein expression in a variety of cell types and tissues [34], many of which are currently available online using World Wide Web (WWW) [35, 36]. We have constructed an IPG 2D-PAGE re-ference map of over a 1000 silver-stained whole tissue lysate polypeptides from normal adult liver and a series of hepatoma cell lines using electro-phoretic conditions similar to those reported by Hochstrasser and co-

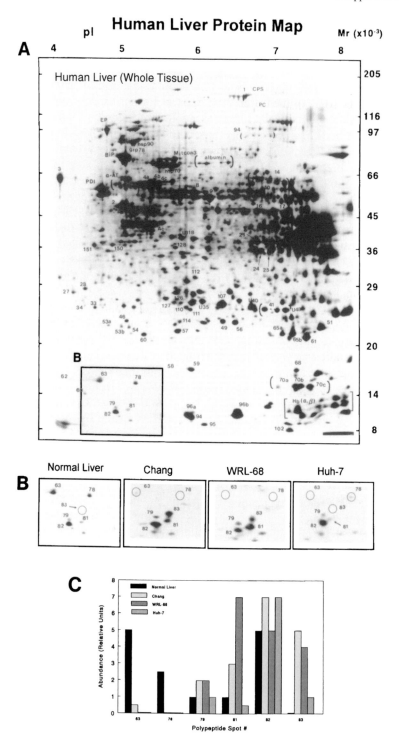

Human Liver Protein Map

workers for the SWISS 2D-PAGE human liver protein database [37–39]. Comparison of the respective IPG 2D-PAGE human liver maps revealed > 80% matching of whole cell lysate proteins. Differences in liver tissue handling procedures as well as inherent heterogeneity of protein expression due to age, sex, and pharmacological variations in the liver samples may account for some of the differences observed, many of which were quantitative in nature [18].

3. Applications

2D-PAGE has found widespread application in a broad range of biological systems that include the analysis of phenotypically dependent alterations of protein expression during normal and abnormal growth and differentiation, such as cancer development; the identification of specific protein changes induced by mutagens and carcinogens, hormone treatment, mitogen stimulation and nutrient changes; as well as for the characterization of human, animal, and plant tissues and cells. Since cellular proteins serve as the "workhorses" to maintain the normal phenotype of the cell, any morphological and/or functional abnormality which occurs as a result of mutations in transformed cells should be reflected at the protein expression level. In ongoing studies, we have been utilizing IPG 2D-PAGE to detect specific polypeptide alterations associated with experimental rodent hepatocarcinogenesis as well as human lung, breast and ovarian carcinogenesis. Figure 2A illustrates an IPG-2D-PAGE map of human liver proteins that includes a representative subset of polypeptides differentially expressed in human hepatomas (Fig. 2B, C). It is anticipated that by identifying proteins from subcellular compartments from different cell types, information concerning both the general and specific biochemical events occurring within the compartments can be discovered.

4. Characterization of proteins from 2D-PAGE

Identification of proteins associated with specific biological conditions and disease states has been based for the most part on their pI, M_r, and relative abundances but has provided no direct information concerning either

---

Figure 2. (A) 2D-PAGE map of human liver and hepatoma proteins. Polypeptides were separated in the first dimension using nonlinear pH 3–10 IPG gel strips. Strips were cut to fit the second dimension SDS/PAGE gel between the pH range of 4 (left) to 8.5 (right). Abscissa, pH range; ordinate, M_r times 10^3. Numbered polypeptides illustrate those that have been identified using either Western immunoblot analysis, protein microsequencing, or comparison with the SWISS 2D-PAGE database of human liver proteins [18, 39]. (B) Representative polypeptides differentially expressed in normal human liver and hepatoma cell lines. (C) Relative abundance of differentially expressed liver proteins.

their biological function or identity. Western immunoblot analysis [40] is the method of choice for the identification of proteins on 2D gels but is severely limited by both the availability of specific antibodies and their capacity to react with the protein(s) of interest on membrane surfaces; thus it is not a general method. However, analytical methods are available for the chemical characterization of proteins such as amino acid composition [41, 42], N-terminal and internal amino acid microsequence analysis [43–46], and mass spectral peptide analysis [47–50].

4.1 Edman N-terminal amino acid sequence analysis

Edman N-terminal amino acid microsequencing affords the most straight-forward method for protein sequence characterization. This methodology fails, however, if the protein is N-terminally blocked. Studies have suggested that up to 50% of all cellular proteins undergo some type of post-translational modification. Internal amino acid sequence analysis overcomes N-terminal blockage but requires chemical or enzymatic cleavage of the blocked protein to generate shorter peptide sequences which must be isolated. In the past, requirements for large sample quantities limited N-terminal sequence analysis of gel-isolated proteins to those most abundantly expressed (i.e. those expressed at $> 10^5$ copies per cell), necessitating time-consuming running of multiple 2D gels with pooling of 20–50 individual polypeptide spots [45, 51–53]. Advances in sequencing technology have pushed "workable" sample quantities to the 0.5–5 pmol level, although sample throughput is still an issue [54]. These limitations have been further minimized with the use of higher loading capacity IPG gel strips in lieu of CA tube gels [17, 18, 51] and has permitted direct N-terminal amino acid microsequence analysis of individual liver proteins from single Ponceau S-stained, IPG 2D-PAGE blotted, PVDF membranes using an initial protein loading of 2 mg [18]. Many of the normal liver and hepatoma proteins gave information on inferred leader sequences since the first sequenced residue was several (20–30) residues from the methionine initiation site predicted by the complementary DNA (cDNA) of the adult liver. Combining sample application and in-gel rehydration of IPG strips allows up to 15 mg of protein to be loaded onto a single gel, and when blotted to PVDF membranes and stained with Amino Black, more than 1000 proteins in high nanogram to low microgram quantities are obtained [55].

   The challenge of sequence database searching is to identify features of proteins which are readily measured, can distinguish one protein from another, and are derivable from the sequences of proteins in the database. Amino acid composition, or the determination of as few as five sequential amino acid residues, can provide identification when compared with genomic sequence databases [56]. Recently, methods relying on mass spectrometers, and advances in sample preparation by *in situ* proteolytic digestion, have made it possible to identify proteins in 2D gels at the subpico-mole level [57–59]. These methods take two approaches. The first, peptide mass fingerprinting, matches the measured masses of proteolytic peptides to those predicted for each protein in a database. The second uses structural information obtained by tandem mass spectrometry to match peptides to known sequences with some degree of interpretation (peptide sequence tag) or without interpretation (fragment ion tags, or a cross correlation algorithm).

4.2 Peptide mass fingerprinting

Peptide maps, as characterized by high pressure liquid chromatography (HPLC) or SDS/PAGE, have long been used as "fingerprints" of proteins for the purpose of comparing two proteins or checking the purity of a known protein. When protein digests are studied by mass spectrometry, a peptide map is generated which can be used to search a sequence database. The more accurately the peptide masses are known, the less chance there is of spurious matches. Therefore, mass spectrometric techniques must achieve both high sensitivity and mass accuracy.

Improvements in three types of mass spectrometers have played a major role in obtaining very accurate mass measurements. The first is a matrix-assisted laser desorption/ionization time of flight mass spectrometer (MALDI-TOF-MS). Two instrumental advances, the use of the ion reflec-tor and delayed extraction, have allowed isotopic resolution of peptides and low parts-per-million mass accuracies [57, 60]. The others are triple qua-drupole (TSQ) and ion trap mass spectrometers (ITMS) with electrospray ionization sources which generate ions from samples in solution. Depend-ing on the rate at which sample is introduced, ESI is roughly divided in-to "normal electrospray" (μl/min to ml/min) "microelectrospray" (50 or 100 nl/min to 1 μl/min), and "nanospray" (low nl/min). Nanospray [59] en-ables analysis of microliter amounts of a sample mixture over a period of 30 min or longer (Fig. 3). TOF, TSQ, and ITMS can all achieve mass ac-curacies better than 0.1 mass units. For rapid screening of peptide masses of tryptic digests, however, MALDI-TOF-MS is becoming the method of choice. This methodology has the advantage of being relatively insensitive to contamination from salts and detergents and has been automated to allow high sample throughput. It is capable of detecting peptides at the low femtomole (fmol) level. Finally, in MALDI-TOF-MS, in contrast to ESI, most peptides carry only one charge and show only one peak in a spectrum which facilitates data interpretation.

Protein identification by peptide mass fingerprinting is widely used for peptide mixtures produced by in-gel tryptic digests; in one case, up to 90% of proteins have been identified by searching sequence databases [61]. A number of programs are available to automatically match peptide mass

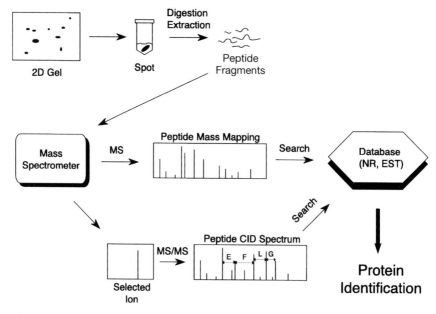

Figure 3. Schematic representation of the experimental appoach used to identify a spot from a 2D-gel analysis. Essentially, a spot of interest is digested with trypsin, and the resulting peptides are analyzed by MS peptide mass fingerprinting and MS/MS peptide sequence tagging. Both results are examined and searched in an appropriate database for the identification of a recognition sequence for the expressed protein.

Table 1. MS Analysis-related Programs and Web sites

PeptideSearch
    EMBL, Protein & Peptide Group
    http://www.mann.embl-heidelberg.de/Default.html

SEQUEST
    University of Washington, Biological Mass Spectrometry Lab
    http://thompson.mbt.washington.edu/sequest/

MS-Fit, MS-Tag, MS-Edman
    UCSF, ProteinProspector
    http://prospector.ucsf.edu/mshome.htm

MassSearch
    ETH, Computational Biochemistry Research Group
    http://cbrg.inf.ethz.ch/subsection3_1_3.html

ProFound, PepFrag
    Rockefeller University, PROWL
    http://prowl.rockefeller.edu/PROWL/prot-id-main.html

Mowse
    Daresbury Laboratory, SEQNET
    http://www.seqnet.dl.ac.uk

Lutefisk97
    Immunex Corp.
    http://www.lsbc.com:70/Lutefisk97.html

---

data against databases (Tab. 1). The program output is usually a list of those proteins in the database that match the input masses; tentative matches are obtained for the most abundant peptides, and the power of peptide mass mapping can be increased by the inclusion of additional structural information. Peptide mass mapping is most useful in analysis of proteins from organisms with completely sequenced genomes, and is of less value in searching expressed sequence tag (EST) databases because of the limited sequence coverage for each entry. Moreover, it is frequently observed that not all of the experimentally determined peptide masses match a list of the possible peptides masses from the protein identified. Several explanations for these nonassigned peptide masses can be entertained; for example, the peptide mass database search might identify a homolog of the protein isolated, some peptides might be nonspecifically cleaved by a protease, or, more interestingly, some peptides might be posttranslationally modified. In such instances, confirmatory criteria should be used to enhance the level of confidence in the protein identification by mass fingerprinting.

4.3 Peptide-sequence and fragment-ion tag

Complete or partial peptide sequences are the most discriminating criteria for the identification of proteins. Such structural information can be obtained from peptide fragmentation data by tandem mass spectrometry (MS/MS), and is better suited for searching the EST database as well as full-length sequence databases (see Fig. 4 for a schematic illustration of the various steps followed for protein identification). In collision-induced dissociation (CID) experiments, a precursor ion collides with an inert gas, and its decomposition products are mass analyzed. Alternatively, in TOF instruments, post-source decay (PSD) can be applied, in which metastable ions decompose during their transit of the flight tube and are resolved by a reflectron [62]. CID is usually performed in TSQ or ion-trap instruments, where the low-energy collision conditions tend to produce rich and interpretable product ion spectra [63]. These two methods have the ability to provide sequence information for femtomolar (or smaller) quantities of peptides; however, since the fragment ion spectra obtained by PSD-MALDI-MS are often difficult to interpret and, in some cases, seem to differ from the spectra obtained by ESI-MS/MS, PSD-MALDI-MS is not used extensively for *de novo* peptide sequencing.

The method of choice for peptide fragmentation and sequencing is currently CID with triple quadrupole or ion-trap instruments. Peptide fragmentation occurs preferentially at amide bonds. Two types of ions appear to be predominant; the (*b*) series, which contains the peptide N-terminus, and the (*y*) series which contains the C terminus. Both type of ions are observed as a ladder of peaks in the mass spectrum whose differences in mass correspond to successive amino acids. Other types of fragmentation

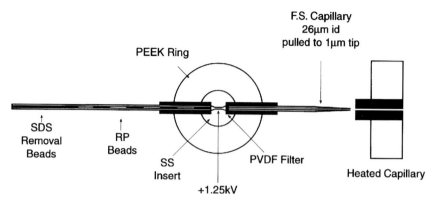

Figure 4. Illustration of a capillary HPLC interfaced with a microelectrospray source. Biphasic columns are packed to effect both SDS removal and reverse-phase chromatography. The columns are coupled to the ESI source through a stainless steel (SS) microvolume union, and at this union voltage is applied through a liquid junction. The ESI source is a fused silica capillary pulled to approximately a 1 μm tip.

can also occur, resulting in a complex spectrum consisting of several overlapping series of ions. The manual interpretation of the CID spectra still remains a challenge and requires some skill and practice. In difficult cases, determining the sequence may require subjecting the sample to selective chemical modification, such as acetylation or methyl esterification. The addition of methyl groups to free side chain groups of aspartic and glutamic acid and terminal α-carboxyl, for example, increases the peptide mass by 14 for each methyl group added and produces mass shifts in the CID spectrum which facilitate interpretation.

Although MS/MS can provide important structural information, precursor ions must be selected in advance. In a nanospray experiment, there is sufficient time during data acquisition to record a full mass range spectrum, select precursors, and perform CID on each ion of interest. Peak widths in LC-MS/MS experiments, however, make MS/MS "on the fly" difficult. Either two experiments must be performed, to select precursors, and then do MS/MS, or a data-dependent experiment can be carried out, in which precursor ions are selected by a computer during data acquisition, and subjected to CID automatically. A fully automated experiment, the "triple play," has been implemented on ion-trap instruments to scan a preset mass range, take a high resolution scan of the most abundant ion, and perform CID on that ion. A variation on this technique is to select precursors based on a MALDI-TOF mass spectrum, and then do data-dependent LC-MS/MS. A routine protocol after eliminating known contaminants such as trypsin or human keratin fragments is to choose prominent ions with masses of 700–2500 Da, which usually are singly charged, from the MALDI mass map, and subject them all to CID in a single LC-MS/MS experiment. Each ion, expected to be doubly charged in the ESI, is then monitored with automated CID, performed as each peptide is detected above an intensity threshold.

In contrast to PSD-MALDI-MS, CID usually produces extended series of ions from which an amino acid sequence can be deduced. However, for the purpose of protein identification, one can in most cases use only two residues derived from three fragment ions or a "Tag" [64] which, when combined with the parent peptide ion mass and the distance in mass units to the N- and C-termini of the peptide, provide a significant amount of additional information. These pieces of information represent a powerful search approach sufficiently specific to identify a protein. Yet, compared with peptide mass fingerprinting by MALDI-TOF-MS, the "peptide sequence Tag" approach is slower and requires some skill in data analysis. The introduction of the SEQUEST program of Yates and co-workers [65, 66] has alleviated some of these problems and opens the way to future possibilities for protein identification (see Tab. 1). The program first identifies all potential peptide fragments which can be generated from a protein in the sequence database whose masses match the experimentally measured masses of the peptide ions. In a subsequent step, it predicts the fragment ions expected for each of the candidate peptides and, using two different algorithms, compares the experimentally determined MS/MS spectrum with the predicted spectra, and the highest-scoring peptide sequences are reported. The success of a search is estimated from the cross-correlation factor reported (X_{corr}), and the difference in X_{corr} between the best and second best matches – higher numbers being better. The data are finally summarized, returning a numerical value for the best-matched peptide and assigns a score for each potential protein match. A consensus match is reported if several searches yielded matches to the same protein. This program has two important advantages: (i) it is tolerant to the analysis of protein mixtures, and (ii) it can be run in a totally automated manner. Searches can be initiated automatically by the mass spectrometer's computer, and it has been very useful for analyzing large number of samples automatically [67]. A further approach, which relies on high-quality tandem mass spectrometry data, has recently been reported [68], with a new algorithm with a FASTA database search engine to generate enough sequence information to perform a similarity analysis of the database.

4.4 Technical challenges and innovations

The approaches described in section 4.3 can only be used when the spectra are of sufficient quality to allow several residues to be assigned without ambiguity. As the analytical systems become more sensitive, five major problems need to be addressed: (i) sample enrichment procedures prior to electrophoresis for the separation of low abundance proteins (e.g. single-copy gene products); (ii) alternative gel matrices that decrease in-gel

protein interactions during electrophoresis; (iii) improved staining procedures to visualize minute quantities of protein in a manner compatible with
subsequent structural characterization; (iv) optimization of the sample
handling with minimal protein losses, and without introducing contami-
nant proteins; (v) improvements in separation systems interfaced directly
to the mass spectrometer, and design of high-sensitivity instruments with
very high mass ranges.

4.4.1 Sample enrichment
A significant degree of enrichment of low-abundance polypeptides prior to
2D-PAGE can be achieved using subcellular fractionation and protein pre-
fractionation. Traditional cell fractionation, utilizing differential size and
sucrose-density centrifugation with subsequent purification of subcellular
organelles would provide a convenient method to obtain highly purified
and enriched organelle-specific protein fractions. Preelectrophoretic enrichment would be expected to reduce the number of proteins observed in
whole cell lysates from thousands to a few hundred for purified organelles.
Subcellular fractionation would be important not only to expand the
"viewing window" of polypeptides catalogued but would also provide
valuable information regarding possible biochemical functions of proteins.
Another approach which shows promising application for organelle enrichment is free-flow and density gradient electrophoresis. These techniques and their applications have recently been reviewed [69] and have been
used successfully for the preparative isolation of single proteins and peptides as well as various subcellular organelles.

Enrichment of protein classes or specific proteins and isoforms may be
achieved by the use of high-affinity binding chromatography. Agarose gels
formulated with immobilized ligands including various dye molecules,
lectins, DNA or RNA structural motifs, or antibodies to specific proteins
could be used to bind and enrich specific proteins. Such an approach has
identified numerous low-abundance tyrosine-phosphorylated proteins from
B lymphocytes using a combination of antiphosphotyrosine affinity column
chromatography, gel electrophoresis, and electrospray mass spectrometry [70].

4.4.2 Alternative gel matrix
Since its introduction in 1959, cross-linked polyacrylamide polymers have
been the matrices of choice for protein electrophoresis [71]. Despite all its
advantages (optical transparency, electrical neutrality, high hydrophilicity, and general inertness to proteins) polyacrylamide still has certain
limitations. Polyacrylamide gels are routinely prepared by a radical-initiat-
ed copolymerization of acrylamide and with a cross-linker, most commonly N,N'-methylenebisacrylamide (BIS), catalyzed by N,N,N',N'-
tetramethylenediamine (TEMED) and ammonium peroxydisulfate (APS).
While a 90% conversion from monomer to polymer occurs within the first

5–15 min, the gel matrix is not homogeneous, displaying random networks of cross-linked strands and pore sizes. However, a more serious concern is that only a small fraction of the BIS molecules become covalently linked, with relatively high concentrations remaining unreacted in the gel matrix. In addition, the polyacrylamide matrix is also susceptible to rapid hydrolysis under both acidic and basic conditions to yield acrylic acid. The resulting free BIS and acrylic acid residues are potentially available for adverse interaction with separating proteins. While some progress has been made in optimizing existing acrylamide-based matrices, most effort has been devoted to the development of more efficient cross-linkers for specific applications. Recently, polyAAEE (N-acryloylaminoethoxyethanol), agarose-acrylamide mixed-bed matrices, and novel polymerization chemistries have been introduced to increase the extent and uniformity of cross-linking during polymerization and to increase polyacrylamide backbone acid/base stability, with the ultimate goal of developing reusable gel matrices [72].

4.4.3 Enhanced staining procedures

While silver staining of proteins on 2D gels is currently the most sensitive of all nonradioactive protein visualization procedures, it is not without problems [12]. Alternative dye staining protocols, including reverse staining with imidazole-SDS-zinc [73], silver-enhanced Coomassie Brilliant Blue, gold/silver staining of blotted proteins, as well as fluorescent stains such as SYPRO Orange and Red [74], all have sensitivity at the femtomolar range (1–10 ng per spot) and offer attractive alternatives to silver staining alone. Each of these procedures can be performed in the absence of or with a minimal "fixative" step, thereby minimizing protein precipitation within the polyacrylamide matrix and facilitating recovery of low-abundance peptides for subsequent mass spectral analysis. In the case of imidazole-SDS-zinc reverse staining, protein elution by passive diffusion was accomplished at room temperature into a detergent-free buffer with high yield (90–98%) at low picomole levels (1 pmol/band) [73].

4.4.4 Optimization of sample recovery

Although subpicomole protein identification is currently achievable, most laboratories still cannot reach such a degree of sensitivity on a routine basis. New methods to increase sample recovery from gels with minimal contamination from interfering substances, like detergents/dyes/stains, and airborne keratins need to be explored in order to limit background contamination, which seriously obscures analysis of nanogram quantities of low-abundance proteins. Reduction in the amount of sample handling at each stage of manipulation, inclusion of a small amount of detergent like SDS to reduce adsorptive losses [75], and cleanup protocols, including final extraction with neat acetonitrile, are essential for routine protein identification at 100 fmol levels. Capillary HPLC utilizing biphasic capillary

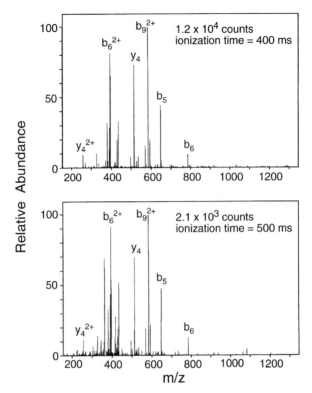

Figure 5. Collision-induced dissociation (CID) spectra of 10 fmol (a) and 250 amol (b) of the [M + 3H]$^{3+}$ ion of the peptide standard angiotensin I injected into a 75 µm HPLC column on-line with a micro-ESI source. The two spectra are virtually idenitical and are of sufficient quality for a database search.

columns to effect both SDS removal and reverse phase chromatography and interfaced directly with microelectrospray ionization will further decrease sample manipulations and associated sample losses. Figure 5 outlines the construct which is interfaced to the mass spectrometer [75].

4.4.5 Improvement in LC-MS interface systems and mass spectral analysis
Improved mass sensitivity for peptide sequence analysis can be achieved using on-line microcapillary HPLC columns to introduce peptides to the mass spectrometer. Because ESI is a concentration-sensitive technique, microcapillary columns operated at sub-µl/min flow rates yield extremely high sensitivity. The development of an on-line gradient LC microspray interface ionization source, constructed using a pulled silica needle coupled to a capillary column through a microvolume union (Fig. 5), has permitted detection of very diluted standards in the low fmol range. As shown in Figure 6, the CID spectrum of 250 attomoles of angiotensin I, generated by the microspray LC-MS/MS, is of sufficient quality for database searches

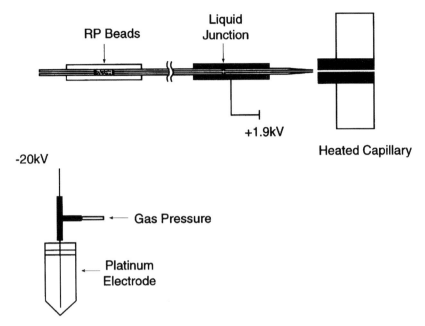

Figure 6. Schematic of the SPE-CE-MicroESI MS/MS system. The first step is to load the sample onto the RP beads of the SPE device by applying pressure at the injection end of the capillary in order to concentrate the sample into a small volume compatible with CE. Once the system has been washed, the peptides are eluted into the CE system, and analysis is started by applying high voltage from the injection end of the capillary to the microelectrospray interface consisting of a liquid junction. The peptides migrating out of the capillary are electrosprayed into the MS by applying a potential of +1.9 kV at the liquid junction.

[75]. Because limits of detection in full mass range scans are determined by chemical noise from the ESI, the practical working level is the low femtomole range. Wilm, Mann and co-workers [59] have introduced a nanoelectrospray ionization source that, when coupled with a single desalting and concentration step before loading the sample, enables analysis of unseparated peptide mixtures at femtomole levels. Moreover, this method gives a stable spray for over 30 min, so that data can be acquired on many peptides in the mixture. In a large-scale project, detection of yeast proteins from 2D gels *via* nanospray ionization was feasible for 100–200 fmol of protein material loaded on the gel. For 49 proteins analyzed, unambiguous search results were obtained [61].

A microscale, variable flow LC-electrospray ionization source has been coupled to the ion-trap mass spectrometer and used to analyze complex mixtures of in-gel protein digests [76]. Utilizing the technique named "peak parking", it has been possible to selectively extend the analysis time over peaks of interest by decreasing the flow rate while the peptide of interest is still contained within the microspray needle. With this system it is possible to perform manual parent ion selection, high-resolution narrow mass range scans, and optimize relative collision energy (RCE) settings for multiple charge states and components present in a single LC peak. This technique, therefore, offers the benefits of microcapillary chromatography along with the extended analysis time offered by the nanospray system. However, even with this "peak parking" approach, analysis of complex mixtures is very demanding.

Among the techniques commonly used for protein and peptide separations, capillary electrophoresis (CE) exhibits the highest sensitivity. Recent advances in capillary sample concentration methods and the use of MS/MS as detectors for CE-separated peptides has permitted the development of capillary electrophoresis-mass spectrometry (CE/MS) interface systems [77, 78]. Among a significant number of possible CE-MS combinations, capillary zone electrophoresis (CZE) is most frequently used in conjunction with MALDI-TOF or ESI-MS. Numerous reports have appeared on the use of MALDI-TOF for the off-line analysis of proteins and peptides separated by CZE. The highest sensitivity has been obtained by directly depositing the eluting analyte onto a cellulose membrane treated with α-cyano-4-hydroxycinnamic acid [79]. The concentration of the sample is in the pmol/l range, which limits the applicability of the technique to digests of relatively abundant proteins. One of the main limitations of the CE-MS technique that has not been fully resolved is the length of time required to produce a mass spectrum. A TOF spectrometer system equipped with an electrospray source and a novel high-speed data acquisition system has been interfaced with CE [80]. A high detection sensitivity (3–6 fmol) was obtained with fused silica columns of 25 micron i.d. (internal diameter) which allowed for fast separation of a mixture of four standard peptides. The recent development of microspray and nanospray sources has also permitted dramatic advances in sensitivity for CE-ESI-MS. The inclusion of a small C18 cartridge on-line with the CZE system permits concentration of analytes contained in a volume of tens of microliters to a few nanoliters within a few minutes; the analytes are then eluted into the CZE capillary in a volume of a few nanoliters [80]. Figure 7 shows a schematic of SPE-CE-Micro ESI MS/MS setup. Several designs have been reported for the CZE-ESI interface with the most successful being based on a liquid junction between the separation capillary and the micro-ESI tip. As shown in Figure 8, low femtomole amounts of protein were successfully identified.

Recently, a combination of the nanoelectrospray ion source, isotopic end labeling of tryptic peptides with ^{18}O, and a new quadrupole/TOF hybrid instrument (QqTOF) have allowed rapid sequencing of small amounts of gel-separated proteins [81] (Fig. 9 outlines the basic instrumentation setup). The mass accuracy obtained was about 0.03 Da, and the mass resolution was in the range of 5000–7000, so the charge state could be unambiguously determined. For example, a peptide from one of the yeast proteins yielded a partial sequence, ETLN with 50 ppm mass accuracy,

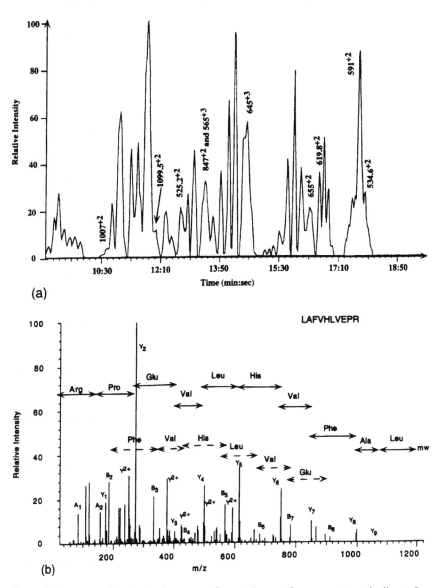

Figure 7. Representative electropherogram of a protein spot from a yeast tryptic digest. See Figure 6 for conditions used for running this sample. (a) Electropherogram indicating *m/z* ratio and charge state of detected peptides; (b) MS/MS spectrum obtained from the *m/z* = 591 peptide ion. (Reproduced with permission from Dr. Ruedi Aebersold).

A) QqTOF

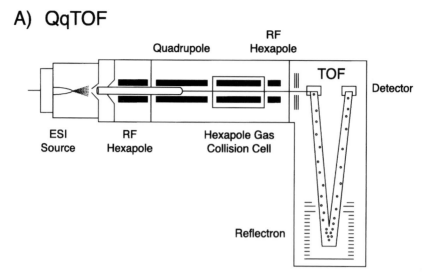

B) FTICR

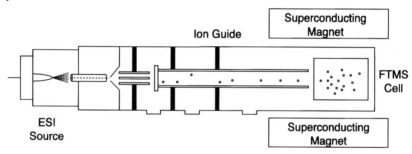

Figure 8. (A) Schematic illustration of QqTOF. In this instrument ions can be mass-selected up to *m/z* 4000 in the first analyzer, induced to undergo collisional fragmentation in the hexapole gas cell, and the fragment ions analyzed using the TOF system. In this instrument good mass resolution and mass accuracy can be achieved with excellent sensitivity. (B) Diagram of the Fourier-transform ion cyclotron resonance mass spectrometer (FTICR). In this instrument the ions are trapped by a combination of electric and magnetic fields in a small cell. These instruments are capable of determining the mass of any single isotope peak to within a few parts per million and obtain structural informations (MS)^n from the same set of ions that were trapped for molecular weight determinations.

which was sufficient to retrieve the peptide sequence from a database and identify the protein of origin despite the occurrence of a posttranslational modification at the amino-terminus of the original peptide. With high mass accuracy and search specificity, this approach may be very useful in searching EST databases and ideal for application of the ^{18}O method of C-terminal labeling [82].

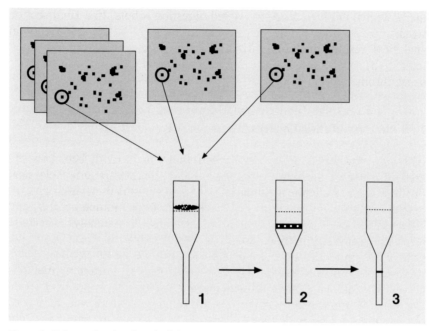

Figure 1. Scheme showing the principle of secondary elution/concentration gel electrophoresis.

gel and the stacking effect guarantee the concentration of eluted proteins in a small volume. Subsequent to the concentration, protein in-gel digestion can take place in a volume as small as 5–10 µl. A relatively large portion of the total available peptide mixture, up to 10%, can then be applied directly for MALDI-MS analysis, and the rest can be separated on RP-HPLC (reverse-phase high performance liquid chromatography) for Edman sequence analysis or MS/MS purposes. Using this approach, we have been able to characterize low picomole to subpicomole amounts of protein starting material employing peptide mass fingerprint analysis [5, 6]. For the characterization of the N-teminus of proteins, the concentrated protein spot can also be electroblotted onto different types of membrane supports suitable for Edman degradation [2], as has been successfully illustrated in the past for the construction of tissue-specific 2D data banks [7].

Recently, a slightly modified version of the elution/concentration technique was published where the standard acrylamide was replaced by a high tensile strength acrylamide [8]; however, results are very similar to those already described earlier [5].

The elution-concentration procedure is nowadays mainly employed for the concentration of minor amounts of proteins and "unknown" proteins (i.e. proteins of which the sequence is not present in a protein or DNA/EST database) from multiple gel pieces prior to RP-HPLC analysis and Edman sequence analysis.

2. Automation in the protein field

2.1. Situation

In contrast to automation in the DNA field, the protein field has long been neglected. This was mainly due to the fact that some years ago the only usable and sensitive final characterization step was Edman degradation, and this reaction scheme has always been a time-limiting factor. With one sequenator, a maximum of one protein per day could be identified (Fig. 2.1). If a protein was N-terminally blocked, and therefore failed to yield any sequence, internal fragmentation of the protein had to be performed (Fig. 2.2). Subsequently, RP-HPLC was used to separate the resulting peptides, and fractions were manually collected (Fig. 2.3). Two or three of these peptides could then be run on an optimized Edman sequencer per day (Fig. 2.4), and were enough to unequivocally identify the protein in a data bank or design oligo primers for further complementary DNA (cDNA)-cloning experiments. The process of internal fragmentation, RP-HPLC, and Edman sequencing took about 1 week.

2.2. Automation in autosampling and fraction collection

Initial attempts to reduce the manual workload associated with this identification scheme were made by the introduction of automated fraction collectors during the RP-HPLC separation (Fig. 2.3) because HPLC is a robust and known process. Several suppliers brought fraction collectors onto the market, but the critical parameter settings needed when working with minor amounts of peptide is still a problem, particularly when peaks elute as a "shoulder" or as a "front". Better peak recognition software must be developed if the use of such devices is to replace the manual collection of RP-HPLC-separated peptides.

The suppliers of robots were successful with autosamplers, which could be coupled to mass spectrometers. Since the introduction of MALDI-MS and electrospray-MS techniques [9, 10], the direct and rapid analysis of peptides and peptide mixtures after digestion became possible, avoiding the laborious Edman sequencing steps (Fig. 2.5) and the RP-HPLC step (Fig. 2.7).

Electrospray mass spectrometry is indeed ideal for automation when coupled to a chromatographic station and an autosampler. The routine analysis of synthetic compounds (peptides or combinatorial products) has had application in the pharmaceutical companies for quite some time now and is one of the fundamentals of pharmacoproteomics. In its simplest form a sample is injected, perhaps after some robotic modification reactions, in a fixed time frame into the MS [11]. In a more complex set-up the autosampler injects into a liquid chromatography system which is coupled to a

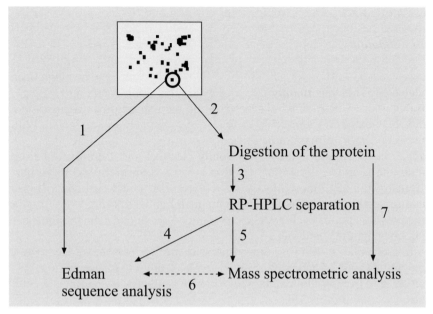

Figure 2. Protein identification scheme.

mass spectrometer and allows mass spectrometers to be in operation 24 h a day. Almost all major MS companies now offer this automation option and allow almost any robot platform and vessel type, even 1536 microtiter plates, to be used.

MALDI-MS is a bit more difficult to automate, because only 0.3 µl or less of sample can be loaded onto a MALDI target, and after drying the spot must be washed with water or a weak acid solution. When working with combinatorial libraries, this requires even several thousands of samples to be loaded and washed in a short time on a small area (of less than 1 cm^2). The samples are either spotted from a robot needle tip or are spotted with a capillary nozzle tool or ink jet set-up. The latter is carried through, e.g. at the MPI for Molecular Genetics in Berlin, by sputtering the sample on a small target size using a system similar to that used in ink jet printers, thus allowing thousands of spots to be loaded on a single small MALDI target. For any type of MALDI target, almost every existing robot platform can be upgraded for this kind of work, and they appear on the market under diverse commercial names. The range goes from cheap and easy-to-use robot set-ups (LabConnections, BAI, Abimed/Gilson) to robots for high-throughput analysis that prepare hundreds to thousands of samples per day (Zymark, Hewlett-Packard, Tecan, Hamilton, Packard, Beckmann). The robot plat-forms can be specifically modified for MALDI purposes, allowing real-time fraction collection of a coupled RP-HPLC, although the coupling of capillary electrophoresis (CE) has also been described [12].

Sample fractionation of digests of down to 500 fmol of protein starting material has been shown at Hewlett Packard [13]. At UCSF (Steve Hall), similar work was done at even smaller levels using Nano-LC at a flow rate between 100 and 200 nl/min and collected directly onto MALDI targets.

2.3. Automation in protein digestion

For proteins already available in protein databanks (at the date of writing more than 312,000), the identification process could be speeded up significantly by using MS techniques. The digest procedure still takes 2 days, but the analysis of the peptide mixture of a known protein now normally requires less than 1 h (Fig. 2.7), including MS analysis, data bank searching and data interpretation.

The most labor-intensive step whether using the Edman or MS methods remains the digest procedure (Fig. 2.2). The protein digestion protocol is a labor-intensive process that is prone to errors and contamination and could benefit from automation.

The first attempts at automating protein digestion steps were made by Hewlett-Packard (HP) [14]. They developed a static work station which is able to manipulate a protein sample immobilized on a biphasic column. Briefly, protein elution, sample washing and desalting, reduction and alkylation are all possible to perform, and the reactor allows immediate loading into an Edman sequencer. CNBr cleavage was demonstrated, and with some manual intervention the elution of peptides from digested protein gel slices onto an RP-HPLC column [13, 15] was performed. The digestion column was thus coupled in front of an RP-HPLC column, allowing the on-line separation of the peptide mixture. The HP-work station is a nice tool for performing simple and single protein manipulation steps, but lacks the ability to process several samples at the same time (e.g. no blank sample can be processed as a control) as well as direct transfer to techniques such as off-line MALDI-MS and NanoES-MS.

Processing of multiple samples, using a principle similar to the above, was introduced by Perseptive Biosystems (now Perkin-Elmer) [16] and BioMolecular Technologies (now with Amersham Pharmacia Biotech) [17]; however, the set-ups are so far only applicable to proteins in solution. The samples are injected and processed in reactor chambers where chemical modifications (e.g. reduction/alkylation), proteolytic digestion using immobilized enzymes (trypsin or Glu-C), RP-HPLC separation and the direct connection to a mass spectrometer, in the so-called LC-MS modus, are all possible. For recombinant proteins available in solution, these work stations offer some advantages, especially when focusing on speed and efficiency of the mapping procedure with the exclusion of manual intervention and (human) contamination.

The instruments mentioned above are important tools for the biotechnology and pharmaceutical industries. However, as the analysis by MS of thousands of proteins separated by 2D gel electrophoresis has gained great importance, the need for the parallel in-gel digestion of multiple proteins is needed. With the exception of the semiautomated HP work station for the digestion of one protein gel spot at a time, no other commercial machine can handle gels or blots. To overcome this problem, we modified and commercialized an Abimed/Gilson robot, applying the digest protocol to up to 30 protein samples in parallel [18]. A special reactor was designed that allowed all reactions to occur in a flow-through manner in less than 10 h and with a sensitivity in the subpicomole range, making the robot an important proteomic tool [19].

Even further automation steps in the protein identification scheme are being attempted for the moment by Oxford Glycosystems (OGS, [20]), where the long-term goal is the characterization of differences in the glyco pattern of different proteins, mainly for diagnostic goals. They implemented an automated spot difference recognition program for 2D gels and linked these data points to feed a robot that then cuts out the spots of interest automatically; a principle already successfully used now for cDNA-colony picking and arraying [21]. These spots should then be digested and analyzed using a robot similar to the one described above or in a microtiter plate. The digestion mixtures are then automatically spotted onto MALDI targets or are injected into an MS using an autosampler coupled to an LC.

These first attempts for a completely automated proteomic laboratory definitely will have to be supported by bioinformatic tools and flexible data handling on one hand and by a good linkage to the biochemical/biological phenomena on the other hand. Overall automation attempts in the protein digestion field will lead to further developments.

3. Peptide sample preparation for MS analysis

3.1 Preparation of biological samples for MALDI-MS analysis

MALDI mass spectrometry has over the past few years evolved to a powerful analytical technique for the characterization of the primary structure of peptides and proteins. Several approaches have been described for this purpose [9], of which peptide mass fingerprinting and post-source-decay (PSD) analysis are mainly employed.

The preparation of MALDI samples seems very simple. However, the preparation is one of the most critical steps in the overall MALDI-MS analysis process. Although it is assumed that MALDI is more tolerant towards the presence of contaminants compared with other ionization techniques like ESI (ES ionization), the incorporation of analytes into growing matrix crystals can be seriously disrupted by these substances. The result is a "bad" sample spot resulting in reduced spectral quality which is reflected by a lower S/N ratio, resolution and sensitivity. However, peptides and proteins can in many cases only be obtained in buffers containing high concentrations of salts, chaotropes and detergents to prevent e.g. their precipitation, increase their solubility or increase the extraction yield of peptides from gel matrices after enzymatic or chemical digestion.

For this reason many different sample preparation techniques have been developed and successfully used for the clean-up of biological samples. Roughly, two different approaches can be distinguished. In the first one, contaminants are removed prior to sample application on the MALDI target. In the second one, contaminants are removed after the sample is spotted on the target, either prior to or after matrix addition.

For the first approach, a manual miniaturized chromatrographic set-up is used. Some researchers have employed RP-HPLC microcolumns packed with different types of RP-chromatographic beads to clean up in-gel digested proteins prior to MALDI-MS [22, 23]. Recently, the use of tips packed with reverse-phase or ion-exchange resins was demonstrated for the removal of salts and detergents from protein digest mixtures [24], and the effectiveness of these tips was illustrated by their recent commercialization [25]. Self-prepared tips containing Poros immobilized trypsin material were also prepared [26], allowing a sample in solution to be passed through and after a while the peptides to be eluted from it. A reduction of autoproteolysis was observed, but the digestion mixture, including buffers, still had to be injected onto an RP-HPLC before the MALDI-MS experiments. In all cases, tips or microcolumns, samples are obtained in small volumes and are essentially free of MALDI contaminants, which results in a possible highly sensitive MALDI-MS analysis of the samples.

For the on-target purification of biological samples, in many cases synthetic membranes or surfaces are employed. These membranes are assembled on top of the MALDI target, and biological mixtures containing peptides and/or proteins are spotted onto the membrane pieces. The strong hydrophobic interactions between the analytes and the membrane surface enable several washing steps to be carried out, and contaminating substances such as buffers, salts and chaotropes are efficiently removed from the target. In most cases, MALDI-matrix solution is subsequently added to the purified analytes, and the sample is ready for mass analysis. Different types of synthetic membranes and surfaces have been successfully used in the past for the analysis of complex biological mixtures. These include active perfluorosulfonated ionomer films [27], polyethylene membranes [28], nonporous polyurethane membranes [29] and C8 and C18 extraction disks [30]. In a recent publication, the use of self-assembled monolayers (SAMs) of octadecyl mercaptan (C18) on gold-sputtered disposable MALDI probe tips was exploited [31]. Here, a hydrophobic surface is created on the target which not only acts as a purification device but at the same time as a sample concentration tool. However, samples have to incubate overnight to become fully concentrated at the probe tip, making the method less feasible for fast usage.

The type of MALDI matrix used, the solvents in which analytes and matrix are dissolved and the manner in which the sample is applied on the target are also key factors for the success of MALDI-MS analysis. For instance, the use of a particular type of matrix, namely 2,5-dihydroxy-benzoic acid, seems to be more tolerant to the presence of certain contaminants since it excludes them during the crystallization process [32]. The use of special prepared thin matrix layers, employing a fast evaporation step, was also demonstrated [33]: it not only improves the sensitivity and resolution in MALDI-MS but also allows extensive washing of samples, effectively removing salts and other impurities.

Although MALDI-MS is intrinsically highly sensitive, as demonstrated by the detection of attomole amounts of peptides [34], the overall sensitivity of any protein or peptide characterization scheme employing MALDI-MS as the analysis step depends on the amount of sample that can be applied as a concentrated spot on a MALDI target. Since only a few microliters of sample can be spotted on a MALDI target, only a very small part of the total available sample (e.g. HPLC eluates) will be analyzed by MALDI-MS. The remainder of the sample will in most cases be discarded since more conventional methods fail to characterize it.

Recently, we developed a procedure to concentrate peptides from highly diluted and/or contaminated solutions such as RP-HPLC eluates and protein digests onto a very small area on a MALDI target, thereby significantly augmenting the sensitivity of MALDI-MS analysis. The procedure "bead-peptide concentration" [35, 36] uses RP-chromatographic beads that are added to diluted and/or contaminated peptide solutions (Fig. 3.1). Through hydrophobic interactions peptides bind very strongly and preferentially to the added beads, whereas contaminating substances such as salts and chaotropes do not [36]. After a short incubation period, peptide-bound beads are harvested by centrifugation, followed by the removal of the supernatant, or the peptide-bead solutions are dried to complete dryness in a SpeedVac concentrator. In both cases a highly concentrated pellet of peptide-bound beads is obtained, which can be transferred to a MALDI target and left to dry. The hydrophobic nature of the added beads makes these beads clustered in a highly concentrated spot (< 1 mm^2) on the MALDI target after drying. Peptides are subsequently on-target eluted by the addition of a small volume of an aqueous/organic MALDI-matrix solution and become at the same time incorporated into the growing matrix crystals (Fig. 3.1).

Using the bead-peptide concentration method, we have demonstrated that femtomole amounts of peptides (RP-HPLC eluates between 10 and 100 fmol) are efficiently concentrated and cleaned up from solutions contaminated with high concentrations of urea, Tris, NaCl and other contaminants [35, 36]. Its speed, simplicity and high concentrating effect

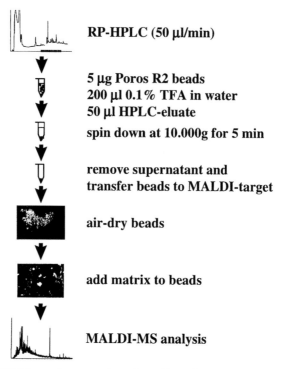

RP-HPLC (50 µl/min)

5 µg Poros R2 beads
200 µl 0.1% TFA in water
50 µl HPLC-eluate

spin down at 10.000g for 5 min

remove supernatant and
transfer beads to MALDI-target

air-dry beads

add matrix to beads

MALDI-MS analysis

Figure 3.1. MALDI sample preparation of peptides using the bead-peptide concentration procedure.

makes the bead-peptide concentration/clean-up procedure a standard step prior to every MALD-MS analysis in the overall protein characterization protocol (Fig. 3.1), allowing the characterization of low femtomole amounts of protein samples loaded on gels [36].

3.2. NanoES-micropurification methods

Although a relatively young technique, nanoelectrospray (NanoES) has been in the field for some years [37]. The central part of the nanoelectrospray source is a borosilicate capillary sputtered with gold to conduct an applied electrical potential to its spraying tip. This set-up allows the sensitive analysis of only minute amounts (femtomole range) of proteins and peptides for 30 min using only 1 µl of sample. Nano-ES-MS/MS can sequence out of a digest mixture up to 10 peptides in 30 min [38].

However, when samples are not clean, salt clustering and other phenomena will disturb the MS-analysis process [39]. Thus the sample must to be cleaned up before analysis. Several approaches that are employed to obtain a clean and concentrated sample for (Nano)ES, and some of the

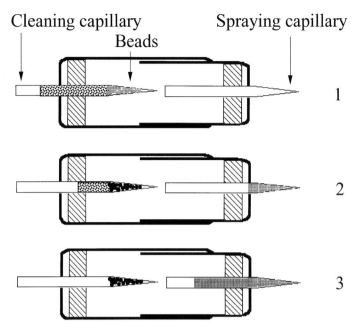

Figure 3.2. Off-line desalting/concentration device used prior to NanoES.

MALDI-MS cleaning-up methods (see 3.1.) can also be used here and vice versa.

One possible method is on-line microdialysis [40]. Protein and peptide solutions having high salt concentrations can be infused into a micro-dialysis tube directly interfaced with a NanoES source. An increased sig-nal-to-noise ratio of a factor of more than 40 is obtainable with this desalting step. However, the method is time-consuming, and sample losses sometimes occur.

Another method to reduce disturbants and increase mass sensitivity is on-line coupling of LC (liquid chromatography) to MS. But as NanoES has no need for pumps and valves, such an on-line coupling would constitute a critical set-up. Nevertheless, on-line dynamic nanoscale liquid chromatog-raphy with a flow of 25 nl/min [41] delivered to an electrospray ionization source using pneumatic splitters [42] has been reported, and will be im-portant in the future for automation purposes.

A third method uses a rapid and robust off-line desalting/concentration step coupled to the NanoES procedure to allow direct analysis of impure samples such as peptide mixtures after in-gel digestion [38]. This set-up (Fig. 3.2) uses two aligned capillaries fixed in a holder, the size of an Eppendorf tube, such that it fits in a table centrifuge.

One capillary contains chromatographic material (Poros perfusion ma-terial), is loaded with the sample mixture, and is centrifuged. The peptides bind to the beads, whereas salts and other MS-disturbing elements can be washed from the column without eluting the peptides (note here that several 2-µl loadings on the cleaning needle can occur, enlarging the sample amounts that can be loaded). Then the spraying needle is placed in its position (Fig. 3.2), and the peptides, sitting on the beads of the cleaning needle, are eluted upon centrifugation with 70% methanol + 1% formic acid. Finally, the spraying needle containing the peptide mixture (Fig. 3.2) is placed in front of the MS. The advantage of this system is that large volumes can be loaded and reduced to only 1 µl or so, and it is possible to step-elute peptides for easy MS data interpretation.

Acknowledgments
K. G. is research assistant of the Fund for Scientific Research-Flanders (Belgium) (F.W.O.). The work presented in this review article was supported by EC grant no. ERB CHRX-CT94-0430 to J.V., by a grant from the Concerted Research Actions (GOA) to J.V., by financial support by the Flemish Community (VLAB-COT no. 035) to J.V. and by the Foundation of Scientific Research-Flanders.

References

 1 Rasmussen H, Van Damme J, Puype M, Gesser B, Celis J, Vandekerckhove J (1991) Micro-sequencing of proteins recorded in human two-dimensional gel protein databases. *Electrophoresis* 12: 873–882
 2 Vandekerckhove J, Rider M, Rasmussen H, De Boeck S, Puype M, Van Damme J, Gesser B, Celis J (1993) Routine amino acid sequencing on 2D-gel separated proteins: a protein elution and concentration gel system. In: K Imahori, F Sakiyama (eds): *Methods in protein sequence analysis.* Plenum Press, New York, 11–19
 3 Rider M, Puype M, Van Damme J, Gevaert K, De Boeck S, D'Alayer J, Rasmussen H, Celis J, Vandekerckhove J (1995) An agarose-based gel-concentration system for microsequence and mass spectrometric characterization of proteins previously purified in polyacrylamide gels starting at low picomole levels. *Eur J Biochem* 230: 258–265
 4 Gevaert K, Rider M, Puype M, Van Damme J, De Boeck S, Vandekerckhove J (1995) New strategies in high sensitivity characterization of proteins separated from 1-D or 2-D gels. In: M Atassi, E Appella (eds): *Methods in protein structure analysis.* Plenum Press, New York, 15–26
 5 Gevaert K, Verschelde J-L, Puype M, Van Damme J, Goethals M, De Boeck S, Vandekerckhove J (1996) Structural analysis and identification of gel-purified proteins, available in the femtomole range, using a novel computer program for peptide sequence assignment, by matrix-assisted laser desorption ionization-reflectron time-of-flight-mass spectrometry. *Electrophoresis* 17: 918–924
 6 Gevaert K, Demol H, Verschelde J-L, Van Damme J, De Boeck S, Vandekerckhove J (1997) Novel techniques for identification and characterization of proteins loaded on gels in femtomole amounts. *J Prot Chem* 16: 335–342
 7 Celis J, Rasmussen H, Gromov P, Olsen E, Madsen P, Leffers H, Honoré B, Dejgaard K, Vorum H, Kristensen D et al (1995) The human keratinocyte two-dimensional gel protein database (update 1995): mapping components of signal transduction pathways. *Electrophoresis* 12: 2177–2240
 8 Kristensen D, Inamatsu M, Yoshizato K (1998) Elution concentration of proteins cut from two-dimensional polyacrylamide gels using Pasteur pipettes. *Electrophoresis* 18: 2078–2084
 9 Karas M (1996) Matrix-assisted laser desorption ionization MS: a progress report. *Biochem Soc Trans* 24: 897–900

3. Examples

3.1. Blank runs

It is of highest importance to perform blank digestions, as well as to make blank runs on the chromatograph. If it is not possible to produce a flat undisturbed baseline as shown in Figure 1, then a cleaning might be necessary. The negative control, i.e. digestion of a blank gel piece usually contains a few peaks, most probably derived from components in the gel and/or staining reagents. The runs in Figure 1 were done on a 1-mm i.d. column.

3.2. In-gel digestion of a novel protein to support cDNA cloning and sequencing

In a collaboration [18] with Britt-Marie Sjöberg, Stockholm University, a preparation of a manganese-containing ribonucleotide reductase from *Corynebacterium ammoniagenes* was analysed. The sample was nearly pure, yet a final purification by SDS/PAGE was performed after reduction and alkylation with iodoacetamide. The two subunits R1E and R2F, of 80 and 50 kDa, respectively, were excised from the gel and digested with endoproteinase LysC, and the peptides obtained were isolated on a 2.1-mm

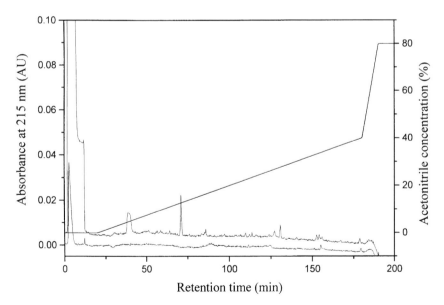

Figure 1. Reversed-phase liquid chromatography on a 1 times 150 mm Kromasil C18 column from Column Engineeering. Upper curve: Negative control, i.e. LysC digestion of a blank gel piece. Lower curve: Chromatographic blank, i.e. no sample was injected. Chromatographic details are given in the text.

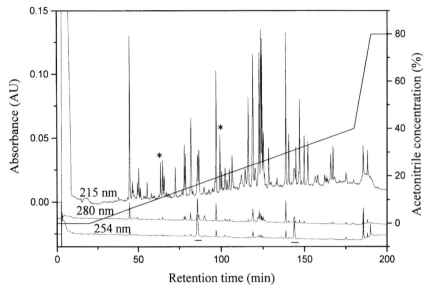

Figure 2. Reversed-phase liquid chromatography on a μRPC C2/C18 SC 2.1/10 column. The sample was the LysC in-gel digest of R1E, the 80-kDa subunit of ribonucleotide reductase from *C. ammoniagenes*. The curves detected at three indicated wavelengths are shown. The material in the fractions labelled with asterisks were Edman-degraded. Underlined fractions represent nonpeptidic contaminants.

i.d. column using the SMART system. Figure 2 shows the peptide map from approximately 15 μg of R1E. Two fractions, labelled with asterisks, were selected for Edman degradation. R2F was treated in the same way (not shown). The two subunits were blotted, in a separate experiment, onto PVDF and N-terminally sequenced, which together with the sequences of the internal fragments confirmed the cDNA sequences.

The detection at the three wavelengths chosen, 215, 254 and 280 nm, reveals some important features of the eluting material. A significant absorption at 280 nm indicates the presence of a Tyr- or Trp-containing peptide. If the corresponding 254-nm trace is approximately half the height of the 280-nm trace, the peptide contains Trp. If the ratio is about 0.25, the peptide contains a Tyr. See Figure 2 for several examples. When the absorption at 254 nm is as high as at 215 nm, as is the case with the two broad, underlined peaks, the peaks do not represent peptides. Peaks with similar characteristics are also found in the negative blank run (see Fig. 1). The peaks in the very last part of the chromatogram also have a high absorbance at 254 nm; they represent the remains of the Coomassie stain.

3.3. In-gel digestion for identity control

During analysis of transport proteins by 2D gel electrophoresis, an un-known protein of approximately 70 kDa was found and was suspected to be involved in the protein complex studied. Since it was present in very small amounts, 36 excised spots were concentrated into a 15% polyacrylamide gel, from which the resulting Coomassie-stained band was excised and analysed by LysC in-gel digestion. Peptide isolation was done on a 1-mm i.d. column (Fig. 3). Although the sample was contaminated with com-ponents having strong absorbance at 254 nm, it was possible to deduce a sequence, RNTTIPTK, from the fraction labelled with an asterisk. This sequence is found only in heat shock proteins, and since the component analysed was 70 kDa, it was concluded that the unknown protein was Hsp 70.

3.4. Comparative peptide mapping by in-gel digestion

During SDS/PAGE analysis one often finds closely eluting bands, although only one band is expected. One reason may of course be that the prepara-tion of the desired protein is heterogeneous or contaminated with a similarly sized protein; alternatively, the sample is in part posttranslationally modi-

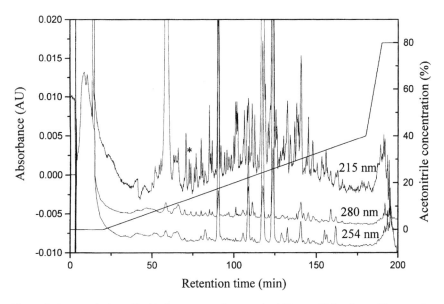

Figure 3. Reversed-phase liquid chromatography on a 1 × 150 mm C18 column. The sample was obtained after concentration of 36 gel spots into a polyacrylamide gel, followed by LysC in-gel digestion. The material under the peak labelled with an asterisk was subjected to Edman degradation.

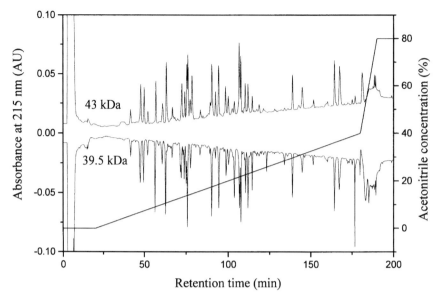

Figure 4. Reversed-phase liquid chromatography on a µRPC C2/C18 SC 2.1/10 column. Two proteins of 43 and 39.5 kDa, respectively, migrating closely in SDS/PAGE, were digested individually with trypsin.

fied, e.g. by glycosylation. By performing separate in-gel digestions of the two closely migrating proteins and comparisons of the resulting peptide maps, it is usually easy to judge whether they are related. Figure 4 depicts a case where two proteins migrating as 43- and 39.5-kDa bands were digested with trypsin in the gel. The peptide maps were almost identical, and there is no doubt that the two proteins are strongly related. We did not analyse the reason for the 3.5-kDa difference, but the heavier protein might be glycosylated.

3.5. In-gel digestion followed by mass spectrometric analysis

Since other chapters deal with several aspects of mass spectrometry, examples of mass spectrometry applications are not given in the present chapter. However, we are routinely analysing in-gel digests by peptide mass mapping. In our experience, when a band is seen only very faintly by Coomassie staining, we get a useful spectrum with MALDI-TOF-MS by analysis of 0.5 µl (about 1%) of the buffer containing the peptide extract. For protein bands visualised by silver staining, we are usually successful after desalting and concentration on the nanocolumn as described above.

4. Discussion

The use of SDS/PAGE as a technique for isolation of proteins for sequence analysis has become widely accepted, especially after the advent of in-gel digestion. This allows an easy way to generate internal protein fragments for Edman degradation from practically any protein. Although SDS/PAGE does not possess the resolving power of 2D electrophoresis, it can generate, when combined with a high purification step, homogeneous protein bands. Hence, using affinity chromatography with an immobilised synthetic peptide as ligand and separating the desorbed material on SDS/PAGE, it is possible to purify and identify Crk proteins from whole cell lysates [19].

The importance of analysis of homogeneous protein bands is obvious. Even a sharp and narrow band in a gel may contain more than one protein. An easy mode of judgement of whether a band contains one or more proteins is to count the number of peptides obtained after chromatography and compare with the expected number, assuming about 5% each of Arg and Lys (for tryptic digestion). i.e. a 50-kDa protein, which has about 440 residues, would yield approximately 45 tryptic peptides. If considerably more peptides are obtained, it is likely that they are derived from more than one protein, and hence that the band is impure.

We have almost exclusively used modified porcine trypsin, sequence grade, or endoproteinase LysC from *Achromobacter lyticus* for in-gel digestion. These two enzymes are very reliable and relatively cheap. As mentioned above, the trypsin will autodigest to a small extent, which gives the possibility of internal calibration during MALDI-TOF-MS. The use of LysC generates longer fragments than use of trypsin, with all fragments except the N-terminal peptide preceeded by a Lys residue. These two characteristics of the trypsin and LysC digests, respectively, make the design of oligonucleotide probes easier.

There has been some controversy in the field whether digestion in-gel or on-membrane is to be preferred. Since the yield of transfer from the gel onto membranes varies greatly between proteins, we prefer the in-gel alternative in order to avoid all unnecessary manipulations when dealing with minute amounts. For very small proteins, however, which may leak out from the gel during washing, a transfer may be advantageous. It was shown, in a recent study, by the Association of Biomolecular Resource Facilities, that the digestion efficiency of what was actually present on the membrane, was comparable to that of in-gel digestion [20].

Acknowledgements
I thank Prof. Carl-Henrik Heldin, of the Ludwig Institute, for helpful discussions, and the following three colleagues for allowing me to use unpublished material as examples: Prof. Britt-Marie Sjöberg, Dept. of Molecular Biology, Stockholm University; Dr. Ulla Lahtinen, Ludwig Institute for Cancer Research, Stockholm Branch; Dr. George Siegenthaler, Dept. of Dermatology, Geneva University Hospital. I also thank Dr. Michael Cross, Dept. of Medical Biochemistry and Microbiology, Uppsala University, for linguistic revision.

References

1 Vandekerckhove J, Bauw G, Puype M, Van Damme J, Van Montagu M (1985) Protein-blotting on Polybrene-coated glass-fiber sheets. A basis for acid hydrolysis and gas-phase sequencing of picomole quantities of protein previously separated on sodium dodecyl sulfate/polyacrylamide gel. *Eur J Biochem* 152: 9–19

2 Matsudaira P (1987) Sequence from picomole quantities of proteins electroblotted onto polyvinylidene difluoride membranes. *J Biol Chem* 262: 10035–10038

3 Aebersold RH, Leavitt J, Saavedra RA, Hood LE, Kent SBH (1987) Internal amino acid sequence analysis of proteins separated by one- or two-dimensional gel electrophoresis after in situ protease digestion on nitrocellulose. *Proc Natl Acad Sci USA* 84: 6970–6974

4 Fernandez J, Andrews L, Mische SM (1994) An improved procedure for enzymatic diges- tion of polyvinylidene difluoride-bound proteins for internal sequence analysis. *Anal Biochem* 218: 112–117

5 Eckerskorn C, Lottspeich F (1989) Internal amino acid sequence analysis of proteins separated by gel electrophoresis after tryptic digestion in polyacrylamide matrix. *Chro- matografia* 28: 92–94

6 Ward LD, Reid GE, Moritz, RL, Simpson RJ (1990) Strategies for internal amino acid sequence analysis of proteins separated by polyacrylamide gel electrophoresis. *J Chromatog* 519: 199–216

7 Rosenfeld J, Capdevielle J, Guillemot JC, Ferrara P (1992) In-gel digestion of proteins for internal sequence analysis after one- or two-dimensional gel electrophoresis. *Anal Biochem* 203: 173–179

8 Hellman U, Wernstedt C, Gonez J, Heldin CH (1995) Improvement of an "in-gel" digestion procedure for the micropurification of internal protein fragments for amino acid sequenc- ing. *Anal Biochem* 224: 451–455

9 Jenö P, Mini T, Moes S, Hintermann E, Horst M (1995) Internal sequences from proteins digested in polyacrylamide gels. *Anal Biochem* 224: 75–82

10 Williams KR, LoPresti M, Stone, K (1997) Internal protein sequencing of SDS/PAGE-se- parated proteins: optimization of an in gel digest protocol. In: DR Marshak (ed) *Techniques in protein chemistry VIII*. Academic Press, San Diego, 79–90

11 Houthaeve T, Gausepohl H, Ashman K, Nillson T, Mann M (1997) Automated protein preparation techniques using a digest robot. *J Protein Chem* 16: 343–348

12 Shevchenko A, Jensen ON, Podtelejnikov AV, Sagliocco F, Wilm M, Vorm O, Mortensen P, Shevchenko A, Boucheri H, Mann M (1996) Linking genome and protome by mass spec- trometry: large-scale identification of yeast proteins from two dimensional gels. *Proc Natl Acad Sci USA* 93: 14440–14445

13 Wessel D, Flügge UI (1984) A method for the quantitative recovery of proteins in dilute solution in the presence of detergents and lipids. *Anal Biochem* 138: 141–143

14 Shevchenko A, Wilm M, Vorm O, Mann M (1996) Mass spectrometric sequencing of proteins silver-stained polyacrylamide gels. *Anal Chem* 68: 850–858

15 Cordoba OL, Linskens SB, Dacci E, Santome JA (1997) "In gel" cleavage with cyanogen bromide for protein internal sequencing. *J Biochem Biophys Methods* 35: 1–10

16 Wilm M, Mann M (1996) Analytical properties of the nanoelectrospray ion source. *Anal Chem* 68: 1–8

17 Annan RS, McNulty D, Huddlestone MJ, Carr SA (1996) International Symposium ABRF '96: Biomolecular Techniques, Abstract S73

18 Fieschi F, Torrents E, Toulokhonova L, Jordan A, Hellman U, Barbe J, Gibert I, Karlsson M, Sjöberg BM (1998) The manganese-containing ribonucleotide reductase of *Corybacterium ammoniagenes* is a class Ib enzyme. *J Biol Chem* 273: 4329–4337

19 Yokote K, Hellman U, Ekman S, Saito Y, Rönnstrand L, Saito Y, Heldin CH, Mori S (1998) Identification of Tyr-762 in the platelet-derived growth factor alpha-receptor as the binding site for Crk proteins. *Oncogene* 16: 1229–1239

20 Williams KR, Hellman U, Kobayashi R, Lane WW, Mische SM, Speicher DW (1997) Inter- nal protein sequencing of SDS/PAGE-separated proteins: a collaborative ABRF study. In: DR Marshak (ed) *Techniques in Protein Chemistry VIII*. Academic Press, San Diego, 99–109

Proteomics in Functional Genomics
ed. by P. Jollès and H. Jörnvall
© 2000 Birkhäuser Verlag Basel/Switzerland

High-resolution two-dimensional gel electrophoresis and protein identification using western blotting and ECL detection

Julio E. Celis and Pavel Gromov

Department of Medical Biochemistry and Danish Centre for Human Genome Research, Aarhus University, Ole Worms Alle, Build. 170, DK-8000 Aarhus C, Denmark

Summary. Two-dimensional gel electrophoresis has been the technique of choice for analyzing the protein composition of cell types, tissues and fluids and is a key technology in modern proteomics projects. Here we describe reproducible procedures for running isoelectric focusing and nonequilibrium pH gradient electrophoresis gels that are based on the carrier ampholyte technology originally described by O'Farrell. Moreover, we present a sensitive immunoblotting procedure that has been used routinely in our laboratory to determine the identity of hundreds of human proteins.

1. Introduction

For the past 20 years, high-resolution two-dimensional polyacrylamide gel electrophoresis (2D-PAGE) has been the technique of choice for analyzing the protein composition of cell types, tissues and fluids, as well as for studying changes in protein expression profiles elicited by various effectors [1–3 and references therein]. The technique, which was originally described by O'Farrell [4, 5] and Klose [6], separates proteins both in terms of their isoelectric point (pI) and molecular weight. Usually, one chooses a condition of interest, for example the addition of a growth factor to a quiescent cell, and lets the cell reveal the global protein behavioral response, as all detected proteins can be analyzed both qualitatively (posttranslational modifications) and quantitatively (relative abundance, coregulated proteins) in relation to each other [1–3, 7–9 and references therein; see also http://biobase.dk/cgi-bin/celis]. Presently, high-resolution 2D-PAGE provides the highest resolution for protein analysis and is a key technique in the emerging area of proteomics research. It belongs together with genomics, complementary DNA (cDNA) arrays, phage antibody libraries and transgenic models to the armamentarium of technology comprising functional genomics [10–12].

Considering the pivotal role of 2D-PAGE in all proteomic projects, we describe here reproducible procedures for running IEF (isoelectric focusing) and NEPHGE (nonequilibrium pH gradient electrophoresis) gels that are based on the carrier ampholyte technology originally described by

O'Farrell [4, 5, 13, see also http://biobase.dk/cgi-bin/celis]. Moreover, we present a sensitive immunoblotting procedure [14, 15] that has been used routinely in our laboratory to determine the identity of hundreds of human proteins, in particular low-abundancy ones.

2. Sample preparation for 2D-PAGE

2.1 Labeling of cultured cells and tissues with [^{35}S]methionine

2.1.1 Solutions

a. *MEM lacking methionine.* Supplemented with antibiotics (100 U/ml penicillin, 100 μg/ml streptomycin) and 10% dialyzed (against 0.9% NaCl) fetal calf serum (FCS). Dispense in 1 ml aliquots and keep at −80°C.

b. *[^{35}S]methionine* (SJ204, Amersham): Aliquot in 100-μCi portions in sterile 1-ml cryotubes. Keep at −20°C. Freeze-dry just before use.

c. *Labeling medium.* Add 0.1 ml of MEM lacking methionine to each ampoule containing 100 μCi of [^{35}S]methionine.

2.1.2 Labeling of cultured cells

1. Plate the cells in a microtiter plate (96 wells) and leave in a humidified, 5% CO_2 incubator (37°C) until they reach the desired density (3000–4000 cells per well).
2. Freeze-dry the [^{35}S]methionine and resuspend in labeling medium at a concentration of 1 mCi/ml. For one well, use 100 μCi of [^{35}S]methionine in 0.1 ml of labeling medium.
3. Remove the medium from the well with the aid of a sterile, drawn-out (under a flame) Pasteur pipette. Wash once with labeling medium. Add the labeling medium containing the radioactivity.
4. Wrap the plate in Saran wrap and place in a 37°C humidified 5% CO_2 incubator for 16 h, or a shorter period if necessary.
5. At the end of the labeling period, remove the medium with the aid of a drawn-out Pasteur pipette. Keep the medium if you would like to analyze the secreted or externalized proteins. Place the 96-well plate at an angle to facilitate removal of the liquid.
6. Resuspend the cells in 0.1 ml of O'Farrell's lysis solution. Pipette up and down (avoid foaming). Keep at −20°C until use.
7. Apply about 1–2 million cpm (hot trichloroacetic precipitable counts) to the first-dimension gels (IEF and NEPHGE) as described below.

2.1.3 Labeling of tissues

Mince the tissues in small pieces (1–2 mm^3) with the aid of a scalpel and label for 14–16 h in a 10-ml sterile plastic conical tube containing 0.2 ml of labeling medium and 50 µCi of [^{35}S]methionine. At the end of the label-ing period the pieces are centrifuged at 2000 g for 2 min, resuspended in 0.3 to 0.4 ml of O'Farrell's lysis solution and homogenized in a small glass homogenizer. Samples are stored at $-20°C$ until use. Apply 20–30 µl of the sample to the first-dimension gels as described below.

3. Sample preparation for immunoblotting

Cultured cells are grown in monolayers in 75-cm^2 flasks until they are near-ly confluent. Thereafter, they are washed twice with Hanks buffered saline and resuspended directly in 1 ml of O'Farrell's lysis solution. To 30–40 µl of this sample we add 5 µl of [^{35}S]methionine-labeled proteins from the same cell line (labeled as described in 2.1.2) to facilitate the identification of the antigen. Blots are first exposed to X-ray films in order to assess the quality of the transfer.

4. 2D-PAGE [4, 5, 13]

4.1 Materials

First-dimension glass tubes (14 cm in length and 2 mm inside diameter; Bie & Berntsen, Denmark) are immersed for at least 30 min in a solution containing 60 ml of ethanol and 40 ml of HCl, and thereafter washed thoroughly with glass distilled water. Spacers are cut from 1-mm-thick polystyrene plates (Metzoplast SB/Hk). The first- and second-dimension chambers as well as the rack to hold the first-dimension tubes are home-made. The aspiration pump for drying the gels can be purchased from HETO (see also http://biobase.dk/cgi-bin/celis).

4.2 Solutions for making first-dimension gels (IEF and NEPHGE)

a. *Lysis solution.* 9.8 M urea (ICN), 2% (w/v) NP-40 (BDH), 2% ampho-lytes pH 7–9 (Pharmacia Biotech), 100 mM DTT (Sigma). Do not heat when dissolving. Aliquot in 2-ml portions and store at $-20°C$.
b. *Overlay solution.* 8 M urea, 1% ampholytes pH 7–9, 5% (w/v) NP-40, and 100 mM DTT. Do not heat when dissolving. Aliquot in 2-ml por-tions and store at $-20°C$.
c. *Equilibration solution.* 0.06 M Tris-HCl, pH 6.8, 2% SDS (Serva), 100 mM DTT, 10% glycerol. Store at room temperature.
d. *Acrylamide solution.* 28.38% (w/v) acrylamide (Bio-Rad) and 1.62% (w/v) *N,N'*-methylenebisacylamide (Bio-Rad). Filter if necessary and store at 4°C.

e. *NP-40.* 10% (w/v) NP-40 in H_2O. Store at room temperature.
f. *Agarose solution.* 0.06 M Tris-HCl, pH 6.8, 2% SDS, 100 mM DTT, 10% glycerol, 1% agarose (Bio-Rad) and 0.002% Bromophenol Blue (Sigma). Dissolve by heating in a microwave oven. Aliquot in 20-ml portions while the solution is still warm and keep at 4°C.
g. *1 M NaOH stock.* Keep at 4°C for not more than 2 weeks.
h. *1M H_3PO_4.* Keep at room temperature.
i. *20 mM NaOH.* Prepare fresh.
j. *10 mM H_3PO_4.* Prepare fresh.

4.3 Solutions for making second-dimension gels (SDS/PAGE)

a. *Solution A.* 30% (w/v) acrylamide; 0.15% (w/v) bisacrylamide. Filter if necessary. Aliquot in 100-ml portions and store at 4°C.
b. *Solution B.* 1.5 M Tris-HCl, pH 8.8. Aliquot in 200-ml portions and store at 4°C.
c. *Solution C.* 1 M Tris-HCl, pH 6.8. Aliquot in 200-ml portions and store at 4°C.
d. *Solution D.* 10% (w/v) acrylamide; 0.5% (w/v) bisacrylamide. Aliquot in 200 ml portions and store at 4°C.
e. *10% SDS.* Filter if necessary and store at room temperature.
f. *10% ammonium persulfate.* This solution should be prepared just before use.
g. *Electrode buffer.* To make 1 l of a 5× solution, weigh 30.3 g of Trizma base, 144 g of glycine and add 50 ml of 10% SDS solution. Complete to 1 l with distilled water and store at room temperature.

4.4 Running first-dimension gels (IEF and NEPHGE)

1. Mark the glass tubes with a line 12.5 cm from the bottom. Seal the bottom end of the tube by wrapping with parafilm and place it standing up in a rack (see also http://biobase.dk/cgi-bin/celis).
2. Mix the urea, H_2O, acrylamide, NP-40 and ampholyte in a tube containing a vacuum outlet. Gently swirl the solution until the urea is dissolved. Do not heat. Add ammonium persulphate and TEMED (N,N,N′,N′-Tetra-methyl-ethylenediamine), mix and degass using a vacuum pump. Use a clean rubber stopper to control the vacuum.
    To make 12 first-dimensional EF gels, use 4.12 g urea, 0.975 ml of acrylamide solution, 1.5 ml of 10% NP-40, 1.5 ml of H_2O, 0.30 ml of carrier ampholytes, pH range 5–7 (Serva), 0.10 ml of carrier ampholytes, pH range 3.5–10 (Pharmacia Biotech), 15 μl of 10% ammonium persulphate and 10 μl of TEMED.

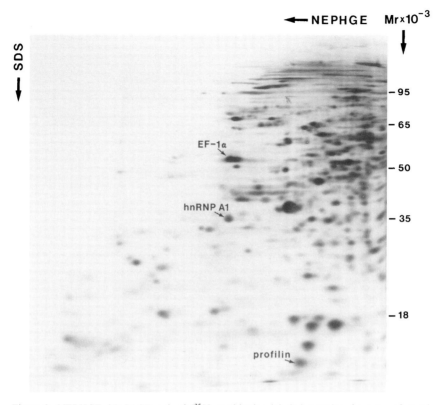

Figure 2. NEPHGE 2D-PAGE gel of [^{35}S] methionine-labeled proteins from transformed human amnion cells. The identity of a few proteins is indicated for reference.

2. Block for 1 h at room temperature or overnight in the cold room.
3. Wash 3 × for 10 min in TBS-Tween or until the washing buffer is clear.
4. Add 10 ml of the primary antibody diluted in TBS-Tween, and incubate for 1.5 h at room temperature.
5. Wash 3 × for 10 min in TBS-Tween.
6. Add 10 ml of peroxidase-conjugated secondary antibody diluted 1 : 1000 in TBS-Tween for 1 h at room temperature.
7. Transfer the membrane blot to a plastic dish and wash 3 × for 10 min in TBS-Tween.

The next steps are carried out in a darkroom:

8. Mix 20 ml of detection solution 1 with 20 ml of detection solution 2 (Amersham kit). This is sufficient for at least eight blots of 13 × 15 cm.
9. Lift the blot carefully and dry it gently by touching a paper towel. Place the blot in a clean plastic dish and add the detection solution. Leave for 1 min (do not shake).

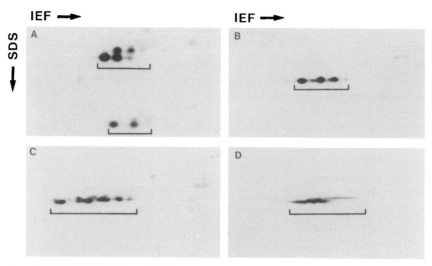

Figure 3. IEF 2D-PAGE blots of whole keratinocyte extracts reacted with antibodies against (A) crk, (B) NCK, (C) MEK2, and (D) p130CAS. Blots were developed using the ECL procedure. Only relevant areas of the gels are shown.

10. Let the solution run of the blot as described earlier, and put it on top of a piece of plastic in a film cassette (protein side up). Cover it quickly with Saran wrap and carefully smooth out air pockets with a piece of paper.
11. Turn off the light (red safety light is allowed), and place an X-ray film on top of the membrane. Close the cassette and expose for 5, 15, 30 s, up to 15 min if necessary.

Figure 3 shows 2D-PAGE blots of whole keratinocyte lysates reacted with antibodies against the low-abundancy proteins crk (Fig. 3A), NCK (Fig. 3B), Mek 2 (Fig. 3C) and p130CAS (Fig. 3D), respectively. To aid the identification of the reacting spots, we usually superimpose the autoradiogram with the ECL film.

6. Remarks

Routinely, about 3000–4000 polypeptides can be resolved (IEF and NEPHGE) and detected using the labeling and running conditions described in this chapter. The lowest level of detection for [^{35}S]methionine proteins correspond to polypeptides that are present in about 40,000 molecules per cell, although immunoblotting in combination with ECL detection may reveal proteins that are present in as few as 1000 molecules per cell. Lower-abundancy proteins may be visualized by increasing the sensitivity of the detection procedure, and/or by the analysis of partially purified cellullar fractions and organelles (see also http://biobase.dk/cgi-bin/celis).

Clearly, a major problem with the carrier ampholyte-based 2D-gel technology has been the lack of reproducibility due to batch variations. In our laboratory, we have overcome this problem by testing various ampholytes (batches, companies) before deciding what combination should be used to carry out a long-term proteomic project. Once a suitable set of ampholytes is found, they are aliquoted in 1-ml portions and frozen until use. In this way, it is possible to run reproducible gels for a long period of time. Today, however, immobilized pH gradients (IPGs) [20–22], which are an integral component of the gel matrix, offer more reproducible focusing patterns, a fact that has made proteomics more attractive to a larger fraction of the scientific community, the pharmaceutical industry included. In our hands, IPGs and carrier ampholytes yield similar separations in broad gradients (http://biobase.dk/cgi-bin/celis), although the IPGs provide better resolution when narrow pH gradients are used.

As mentioned in the introduction, 2D-PAGE plays a key role in proteomics, as it provides the foundation for building up comprehensive protein databases that aim at linking protein and DNA mapping and sequencing information from genome projects (http://biobase.dk/cgi-bin/celis; http://expasy.hcuge.ch/ch2d/2d-index.html). The establishment of such databases has been greatly stimulated by the continous development of sensitive techniques (immunoblotting [14], Edman degradation peptide sequence analysis [23, 24], mass spectrometry [25–28]) for rapidly assessing the identity of proteins (see also other chapters in this book). In our laboratory, we have used a variety of protein identification techniques to build large proteomic databases (human keratinocytes, bladder transitional cell carcinomas, squamous cell carcinomas, urine; http://biobase.dk/cgi-bin/celis), but in particular, 2D-PAGE immunoblotting has been most valuable for the identification of very low abundancy proteins as well as for revealing their modified variants (Fig. 3). In the future, 2D-PAGE proteomic databases will not only annotate genomes but will also provide a global approach to the study of gene expression both in health and disease. One aim of our studies is to use proteomic technology to identify signaling pathways and components that are affected in bladder cancer progression [29, 30], and that may provide new targets for drug discovery.

Presently, there are many problems associated with the 2D-PAGE technology that need to be addressed in the near future. These include the detection of very low abundancy proteins, the separation of very basic polypeptides, as well as the lack of suitable quantitation procedures to analyze all of the proteins resolved in a gel. In addition, it is important to keep in mind that results obtained with the 2D-PAGE gel technology may not be easy to interpret, in particular when this powerful technology is applied to the study of normal and diseased tissue biopsies, which are often quite heterogeneous.

Acknowledgments
We would like to thank I. Andersen, B. Basse, A. Celis, J.B. Lauridsen and G. Ratz for expert technical assistance. The work was supported by grants from the Danish Biotechnology Programme, the Danish Cancer Society and the Danish Centre for Molecular Gerontology.

References

1 Celis JE, Bravo J (eds) (1984) *Two-Dimensional Gel Electrophoresis of Proteins: Methods and Applications*. Academic Press, New York

2 Celis JE, Rasmussen HH, Leffers H, Madsen P, Honore B, Gesser B, Dejgaard K, Vande-kerckhove J (1991) Human cellular protein patterns and their link to genome DNA sequence data: usefulness of two-dimensional gel electrophoresis and microsequencing. *FASEB J* 5: 2200–2208

3 Wilkins MN, Sanches JC, Gooley AA, Appel RD, Humphery-Smith I, Hochstrasser DF, Williams KL (1996) Progress with proteome projects: why all proteins expressed by a genome should be identified and how to do it. *Biotech Gene Eng Rev* 13: 19–50

4 O'Farrell PH (1975) High resolution two-dimensional electrophoresis of proteins. *J Biol Chem* 250: 4007–4021

5 Klose J (1975) Protein mapping by combined isoelectric focusing and electrophoresis of mouse tissues. A novel approach to testing for induced point mutations in mammals. *Humangenetik* 26: 231–243

6 O'Farrell PZ, Goodman HM, O'Farrell P (1977) High resolution two-dimensional electrophoresis of basic as well as acidic proteins. *Cell* 12: 1133–1141

7 Lottspeich F (ed) (1996) High resolution two-dimensional electrophoresis. *Electrophoresis* 5: 811–966

8 Celis JE (ed) (1996) Special Issue: Electrophoresis in Cell Biology. *Electrophoresis* 11: 1655–1797

9 Celis JE, Gromov P, Østergaard M, Madsen P, Honore B, Dejgaard K, Olsen E, Vorum H, Kristensen DB, Gromova I et al (1996) Human 2-D PAGE databases for proteome analysis in health and disease: http://biobase.dk/cgi-bin/celis. *FEBS Lett* 398: 129–134

10 Celis JE, Østergaard M, Jensen NA, Gromova I, Rasmussen HH, Gromov PS (1998) Human and mouse proteomic databases: novel resources in the protein universe. *FEBS Lett:* 430, 64–72

11 Miklos GL, Rubin, GM (1996) The role of the genome project in determining gene function: insights from model organisms. *Cell* 86: 521–529

12 Oliver SG (1996) From DNA sequence to biological function. *Nature* 379: 597–600

13 Celis JE, Ratz G, Basse B, Lauridsen JB, Celis A, Jensen NA, Gromov P (1997) High-resolution two-dimensional gel electrophoresis of proteins: isoelectroc focusing (IEF) and nonequilibrium pH gradient electrophoresis (NEPHGE). In: JE Celis, N Carter, T Hunter, D Shotton, K Simons, JV Small (eds): *Cell Biology: A Laboratory Handbook*. Academic Press, Vol. 4, 375–385

14 Towbin H, Staehelin T, Gordon J (1979) Electroforetic transfer of proteins from polyacryl-amide gels to nitrocellulose sheets: procedure and some applications. *Proc Natl Acad Sci USA* 76: 4350–4354

15 Celis JE, Lauridsen JB, Basse B (1997) Determination of antibody specificity by Western blotting and immunoprecipitation. In: JE Celis, N Carter, T Hunter, D Shotton, K Simons, JV Small (eds): *Cell Biology: A Laboratory Handbook*. Academic Press, Vol. 4, 429–437

16 Laemmli UK (1970) Cleavage of structural proteins during the assembly of the head of bacteriophage T4. *Nature* 227: 680–685

17 Celis JE, Olsen E (1997) One-dimensional sodium dodecyl sulfate-polyacrilamide gel electrophoresis. In: JE Celis, N Carter, T Hunter, D Shotton, K Simons, JV Small (eds): *Cell Biology: A Laboratory Handbook*. Academic Press, Vol. 4, 361–370

18 Laskey RA, Mills AD (1975) Quantitative film detection of ^{3}H and ^{14}C in polyacrylamide gels by fluorography. *Eur J Biochem* 56: 335–341

19 Merril CR, Creed GJ, Allen RC (1997) Ultrasensitive silver-based stains for protein detection. In: JE Celis, N Carter, T Hunter, D Shotton, K Simons, JV Small (eds): *Cell Biology: A Laboratory Handbook*. Academic Press, Vol. 4, 421–428

20 Bjellqvist B, Ek K, Righetti PG, Gianazza E, Görg A, Westermeier R, Postel W (1982) Isoelectric focusing in immobilized pH gradients: principle, methodology, and some applications. *J Biochem Biophys Methods* 6: 317–339

21 Görg A, Pistel W, Günther S (1988) The current state of two-dimensional electrophoresis with immobilized pH gradients. *Electrophoresis* 9: 531–546

22 Righetti PG (1990) *Immobolized pH Gradients: Theory and Methodology*. Elsevier, Amsterdam

23 Vandekerckhove J, Bauw G, Puype M, Van Damme J, Van Montagu M (1985) Protein-blotting on polybrene-treated glass-fiber sheets: a basis of acid hydrolysis and gas phase sequencing of picomole quantities of protein previously separated on SDS-polyacrylamide gel. *Eur J Biochem* 152: 9–19

24 Aebersold RH, Teplow DB, Hood LE, Kent SB (1986) Electroblotting onto activated glass: high efficiency preparation of proteins from analytical sodium dodecyl sulfate polyacrylamide gels for direct sequence analysis. *J Biol Chem* 261: 4229–4238

25 Patterson SD, Thomas D, Bradshaw RA (1996) Application of combined mass spectrometry and partial amino acid sequence to the identification of gel-separated proteins. *Electrophoresis* 17: 877–891

26 Mann M (1996) A shortcut to interesting human genes: peptide sequence tags, expressed-sequence tags, and computers. *Trends Biochem Sci* 21: 494–495

27 Pappin DJ (1997) Peptide mass fingerprinting using MALDI-TOF mass spectrometry. *Methods Mol Biol* 64: 165–173

28 Roepstorff P (1997) Mass spectrometry in protein studies from genome to function. *Curr Opin Biotechnol* 8: 6–13

29 Celis JE, Østergaard M, Basse B, Celis A, Lauridsen JB, Ratz GP, Andersen I, Hein B, Wolf H, Ørntoft TF et al (1996) Loss of adipocyte-type fatty acid binding protein and other protein biomarkers is associated with progression of human bladder transitional cell carcinomas. *Cancer Res* 56: 4782–4879

30 Østergaard M, Rasmussen HH, Nielsen, HV, Vorum H, Ørntoft TF, Wolf H, Celis JE (1997) Proteome profiling of bladder squamous cell carcinomas: identification of markers that define their degree of differentiation. *Cancer Res* 57: 4111–4117

Proteomics in Functional Genomics
ed. by P. Jollès and H. Jörnvall
© 2000 Birkhäuser Verlag Basel/Switzerland

Nanospray mass spectrometry in protein and peptide chemistry

William J. Griffiths

Department of Medical Biochemistry and Biophysics, Karolinska Institutet,
SE-171 77 Stockholm, Sweden

Summary. The introduction of electrospray in the mid-1980s revolutionised biological mass spectrometry, in particular in the field of protein and peptide sequence analysis. Electrospray is a concentration-dependent, rather than a mass-dependent process, and maximum sensitivity is achieved at low flow rates with high-concentration, low-volume samples. This has lead to the development of nanoelectrospray, microelectrospray and related low flow-rate forms of electrospray which offer high sensitivity with low sample consumption. In this chapter the physical chemistry of low flow-rate electrospray is discussed, and a brief review of the types of low flow-rate electrospray interfaces is made. An indication of the performance obtainable on various instruments is given, along with some comments from the author's own experience of these techniques.

Introduction

Mass spectrometry determines the mass to charge ratio (m/z) of gas phase ions. The earliest studies were performed by J. J. Thomson [1] and his student F. W. Aston [2] around the turn of the century. Throughout the 20th century mass spectrometry has played an important role in analytical chemistry, but it is only in the last 30 years that mass spectrometry has become routinely applied to peptide and protein biochemistry. The reason for the late breakthrough of mass spectrometry into protein chemistry was the absence until 1981 of a user-friendly method for ionisation and vaporisation of peptides and proteins (without, in the process, causing their decomposition), making them accessible for mass analysis. However, with the advent of fast atom bombardment (FAB) ionisation [3], peptides and small proteins could be readily analysed by mass spectrometry. In 1988, two other ionisation methods for the mass spectrometric analysis of proteins were reported. At the 11th International Mass Spectrometry Conference in Bordeaux, Karas and Hillenkamp described the ultraviolet laser desorption of ions of masses above 10 kDa, while at the 36th ASMS Conference in San Francisco, Fenn and co-workers [4] presented data on the electrospray (ES) mass spectrometry of proteins. These two new ionisation methods have revolutionised biological mass spectrometry. Karas and Hillenkamp's [5] laser desorption technique, now called matrix assisted laser desorption ionisation (MALDI), when interfaced to time-of-flight (TOF) mass spectrometers has become well established in protein chemistry laboratories, as has ES when fitted to quadrupole mass spectrometers.

The electrospray process: the early years, Dole to Fenn

In the ES process a sample solution is sprayed from a fine needle (or capillary), raised to a potential of a few kV, into a bath gas at atmospheric pressure. The high electric field at the needle tip charges the surface of the emerging liquid, dispersing it into a fine spray (Fig. 1). If the needle is raised to a high positive potential, positively charged droplets are formed, from which solvent evaporates, leading to smaller, highly charged droplets, which when the Rayleigh limit is reached (coulombic repulsion equals surface tension) break up into smaller droplets, which themselves repeat this process until very small droplets result. The mechanism for the formation of gas phase ions from these very small droplets has not been exactly determined, although two plausible models exist (discussed below). The resulting gas phase ions with accompanying bath gas flow through a sampling cone to emerge as a supersonic free jet into a vacuum chamber. A portion of the free jet passes through a skimmer to a second vacuum chamber leading to the mass analyser. The earliest attempts to use an ES dispersion of an analyte solution in a bath gas to produce ions for mass analysis were carried out by Dole and co-workers [6]. Dole reasoned that an ES dispersion of a sample solution would result in the production of

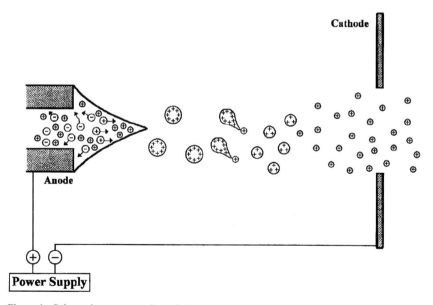

Figure 1. Schematic representation of the electrospray process. A high positive potential is applied to the capillary (anode), causing positive ions in solution to drift towards the meniscus. Destabilisation of the meniscus occurs, leading to the formation of a cone and a fine jet emitting droplets with excess positive charge. Gas phase ions are formed from charged droplets in a series of solvent evaporation-Coulomb fission cycles. (Modified from [12]).

ultimate droplets containing only one charge and one solute molecule. As the last solvent evaporated from such a droplet, its charge would be retained by the solute molecule, thus giving rise to an ion. This model for gas phase ion formation is known as the charge residue model (CRM). Although Dole and co-workers observed ions which were mass-analysed by a retarding potential method, it appears that their belief that they had produced singly charged macromolecules was incorrect. The first successful ES mass spectrometry experiments were conducted by Yamashita and Fenn [7, 8]. They coupled an ES interface to a quadrupole mass filter of m/z range of up to 400, and with it were able to analyse singly charged ions from many different classes of small polar molecules. The range of the mass analyser was later extended to m/z 1500, and with this apparatus Fenn and co-workers started to analyse peptides and proteins [4, 9]. When proteins were analysed by ES mass spectrometry [4, 9], they were found to produce multiply protonated molecules, and give spectra consisting of a sequence of coherent peaks, which differed from their adjacent neighbours by a single charge. It was found that the resulting envelopes of peaks tended to fall in the m/z window 500–1500 for a great variety of proteins ranging from bovine insulin (5730 Da) to chicken egg conalbumin (76 kDa). Fenn and co-workers [9] preferred Iribarne and Thomson's [10, 11] ion evaporation model (IEM) to Dole's charge residue model to explain the formation of gas phase ions from small, highly charged droplets. The IEM differs from the CRM in that it predicts that for very small droplets (R $\leq$ 10 nm) [12] direct ion emission from the droplets becomes dominant over Coulomb fission. Both models explain the production of gas phase ions from very small charged droplets but differ in the predicted size of the droplet from which the gas phase ion is formed.

1. Physical chemistry of low flow-rate electrospray

ES is a concentration-rather than mass-dependent process; thus improved sensitivity is obtained for high-concentration, low-volume samples. This can be explained by considering the physical chemistry of the process. The ES interface itself is a special form of electrochemical cell [12]. When operated in the positive-ion mode, the spray capillary (or needle) is raised to a high positive potential (0.5–4 kV) relative to the rest of the ion source and becomes the anode, whereas the sampling cone becomes the cathode or counter electrode (Fig. 1). The orifice of the spray capillary may be of the order of 0.1 mm (i.d.) for conventional ES, and 0.5 µm (i.d.) for low flow-rate ES. The electric field (E_c) generated at the capillary tip is very high (10^6 V/m), and can be approximated from eq. (1):

$$E_c = 2V_c/[r_c \ln (4d/r_c)] \tag{1}$$

where, V_c is the applied potential (0.5–4 kV), r_c is the capillary outer radius (10 μm–10^{-4} m), and d is the distance from the capillary tip to the counter electrode (0.002–0.02 m). The first and second values given in brackets represent values appropriate for low flow-rate and conventional ES, respectively. From eq. (1) it can be seen that E_c is proportional to V_c, and that E_c is essentially inversely proportional to r_c, whereas E_c decreases slowly with the electrode separation d as a result of its logarithmic dependence. As a consequence of the high field at the capillary tip, positive and negative ions in the electrolyte solution will move until a charge distribution results which counteracts the imposed field. When the capillary is positive, positive ions drift downfield towards the liquid meniscus and negative ions away from it (Fig. 1). Repulsions between positive ions at the surface overcome the surface tension of the liquid, and the surface expands, allowing positive charges and liquid to move further downfield. A cone forms, i.e. the "Taylor cone" [13], and a fine jet, emerging from the cone tip, breaks up into small charged droplets [12]. The droplets are positively charged as a result of excess positive electrolyte being present at the surface of the cone and jet cone. The charged droplets drift downfield through the bath gas towards the counter electrode, shrinking due to solvent evaporation along the way. Either the CRM or IEM can be used to explain the subsequent formation of gas phase ions from the resultant small droplets. With the continual emission of positively charged droplets from the capillary, to maintain charge balance, oxidation occurs within the capillary. If the capillary is metal, reaction (2) may occur at the liquid-metal interface:

$$M(s) \rightarrow M^{2+}(aq) + 2e^- \text{ (in metal)} \tag{2}$$

or alternatively negative ions may be removed from solution by electrochemical oxidation, as in reaction (3).

$$4\,OH^-(aq) \rightarrow O_2(g) + 2\,H_2O + 4e^- \text{ (in metal)} \tag{3}$$

An important factor in determining the efficiency of the formation of gas phase ions via the ES process is the size of the initial, charged droplets emitted from the spray capillary. Pfeifer and Hendricks [14] derived eq. (4), which relates droplet radius (R), to flow rate (V_f), and the electric field at the capillary tip (E_c).

$$R = [(3\varepsilon\gamma^{1/2}\,V_f)/(4\pi\,\varepsilon_0^{1/2}\,KE_c)]^{2/7} \tag{4}$$

where, γ is the surface tension of solvent, ε is the permittivity of the solvent, ε_0 is the permittivity of vacuum, and K is the conductivity of the solution. Although the equation was derived on the basis of unproven assumptions, the prediction of the radius agrees approximately with experiments [12], and also with a semiempirical derivatisation by Fernandez de la Mora and Locertales [15], eq. (5).

$$R = (V_f \, \varepsilon / K)^{1/3} \tag{5}$$

Both equations indicate that at lower flow rates smaller droplets are formed. A consequence of the formation of smaller droplets via low flow-rate ES is an improvement in the efficiency of gas phase ion formation. This can be perceived by consideration of the droplet-evaporation process, where gas phase ions are formed from charged droplets in a series of solvent evaporation-Coulomb fission cycles. As these processes must occur within a given time scale (defined by the geometry of the interface), the smaller the initial droplet the greater the chance of the formation of very small droplets from which gas phase ions are formed. A second advantage of low flow-rate ES over conventional ES is the more efficient sampling of gas phase ions in the low flow-rate interface as a consequence of the reduced spatial dispersion of charged droplets.

A further factor to consider in the efficiency of gas phase ion formation is the hydrophobicity of the ion itself. The more hydrophobic an ion, the greater its surface activity, and the higher the probability that it will sit near the surface of a charged droplet and subsequently become a gas phase ion. This is important for multicomponent systems when considering the relative mass spectrometric sensitivities of the component species. Tang and Kebarle [16, 17] proposed that for a multicomponent system with three electrolytes A, B and C, the mass spectrometric current for one of the electrolytes, i.e. I_A for ion A^+ can be expressed by eq. (6):

$$I_A = f \, p \, I_0 \, \{k_A[A^+]/(k_A[A^+] + k_B[B^+] + k_c[C^+])\} \tag{6}$$

where, I_0 is the capillary current, $[A^+]$, $[B^+]$ and $[C^+]$ are the concentrations of the electrolytes in the solution to be electrosprayed, k_A, k_B and k_C are rate constants for the transfer of the respective ions from the droplets to the gas phase, f is the fraction of charges on the droplets that are converted to gas phase ions and p is the ion sampling efficiency of the mass spectrometer. The greater the surface activity of an ion, the greater the rate constant k and the greater its mass spectrometric current. Also, the smaller the droplet, the greater the fraction of charges converted to gas phase ions, i.e. the greater the value of f. At low flow rates negative effects on sampling efficiency are minimised, resulting in values for p close to unity as compared with 10^{-4} for conventional ES interfaces.

2. Low flow-rate electrospray mass spectrometry

Some of the earliest work on low flow-rate ES was performed by Gale and Smith [18]. They constructed a pressure-infusion low-flow rate ES interface which was coupled to their quadrupole mass spectrometer. The most important feature of this interface was the etched-tip fused silica spray capillary. Capillaries of different i.d. (5, 10, 20 µm) etched in hydrofluoric acid and manually trimmed were tested. Electrical contact was made through a liquid/liquid junction, and a coaxial sheath gas of SF_6 employed to minimise electrical discharge. Gale and Smith found that they could position their spray capillary closer to the sampling cone than for conventional ES, and that the optimum voltage required to maintain a stable ES was only 2.5 kV as opposed to 4 kV with their conventional ES interface. Using a 20 µm i.d. capillary and a flow rate of 0.2 µl/min, they found an improvement in sensitivity for low flow-rate ES over conventional ES of a factor of about 20. By reducing the capillary i.d. to 5 µm and the flow rate to 50 nl/min, further improvements in sensitivity were noted. Several proteins and oligonucleotides (negative-ion mode) were analysed, and sensitivities in the high attomole to low femtomole range were reported. Improved spray stability was also noted.

3. Micro-electrospray mass spectrometry

Following on from Gale and Smith's work, Emmett and Caprioli [19] developed a modified form of low flow-rate electrospray called micro-electrospray. Emmett and Caprioli's micro-ES was essentially similar to Gale and Smith's low flow-rate ES. The spray capillaries (i.d. 5–250 µm) were either etched in hydrofluoric acid and manually ground, or just manually ground, to give a flat-ended, tapered, needlelike tip. Flow rates generated by pressure infusion were 0.3–0.8 µl/min, and capillary voltages were 2.0–3.5 kV when their interface was coupled to a quadrupole mass spectrometer. Emmett and Caprioli [19] found that micro-ES operates best with 50 µm capillaries which were found to give stable, trouble-free sprays. Spectra of myoglobin, carbonic anhydrase and bovine serum albumin were obtained from tens of femtomoles of sample. The major advance to low flow rate-ES made by Emmett and Caprioli was to couple it with liquid chromatography (LC). Driven by a desire to analyse low concentration neurosubstances, they decided to pack unetched capillaries (i.d. 50 µm) with C-18 particles. The resulting column could act as a preconcentration/desalting device. Using this column, effective desalting was achieved, e.g. 50 fmol of methionine enkephalin being detected in 10 µl of Ringer's solution (5 mM KCl, 120 mM NaCl, 1.2 mM $MgCl_2$, 1.8 mM $CaCl_2$, 0.15% phosphate-buffered saline, pH 7.4). Also, by using the column as a preconcentration device, high sensitivities were achieved, i.e. 1 fmol of methionine enkephaline was detected in a full mass scan, and 500 amol in a short scan tandem mass spectrometry (MS/MS) experiment. In subsequent work by Andren and co-workers [20], sensitivity was pushed further to the zeptomole level. It is worth noting that the name "micro-ES" refers to the microbore size (i.d.) of the spray capillary and has nothing to do with the flow rates at which the device is operated or the size of the charged droplets formed. Recently, Emmett and co-workers [21] interfaced micro-ES/LC to a Fourier transform ion cyclotron resonance (FTICR) mass spectrometer. Using micro-ES with capillaries (i.d. 25 μm) ground and packed with C-18 particles, LC/MS experiments were performed at sub-μl/min flow rates giving sensitivities for small peptides of the order of 5 fmol in full mass spectra at an instrument resolving power of greater than 5000.

4. Nanoelectrospray mass spectrometry

Remarkably, at almost the same time as Emmett and Caprioli [19] submitted their original micro-ES paper (received by the *Journal of the American Society for Mass Spectrometry* on 14 March, 1994), Wilm and Mann [22] submitted a paper for publication also on a form of micro-ES (received by the *International Journal of Mass Spectrometry and Ion Processes* on 13 March, 1994). Wilm and Mann [23] named their form of ES nanoelectrospray (nano-ES). The major difference between Wilm and Mann's nano-ES and the other forms of low flow-rate ES discussed above, is that nano-ES is a purely electrostatic process and does not require pressure infusion or nebuliser gas flow. The name nanoelectrospray is intended to reflect the low flow (nl/min) and droplet size (200 nm) characteristics of the interface. Wilm and Mann were stimulated to build their nano-ES interface by the realisation that small droplets with a high surface-to-volume ratio are more likely to efficiently desorb large ions than large droplets. They developed a theoretical model for the electrostatic dispersion of liquids in electrospray, which they qualitatively verified experimentally [22]. Their model predicts proportionality between the two-thirds power of flow rate (V_f) and the size of droplets (R) emitted from the tip of a stable Taylor cone.

$$R = \{\rho/4\pi^2 \gamma \tan(\pi/2 - \upsilon)[(V_a/V_T)^2 - 1]\}^{1/3} V_f^{2/3} \qquad (7)$$

where, ρ is the density of the liquid, γ is the surface tension of the liquid, υ is 49.3° for the classical Taylor cone, V_a is the applied voltage and V_T is the threshold voltage. Earlier equations developed by Pfeifer and Hendricks [14] (eq. 4) and Fernandez de la Mora and Locertales [15] (eq. 5) were described, which like eq. 7 above predict a decrease in droplet radius (R) as flow rate is reduced. Thus, Wilm and Mann reasoned that a more efficient ES interface would be one operating at low flow rates where small droplets would be generated which would efficiently desorb large ions. To this end they built an interface where sample (1–2 μl) is loaded into a gold-coated borosilicate needle (capillary) which is tapered to give a spraying orifice of 1–2 μm, and is positioned 1–2 mm in front of the sampling cone. When the interface was coupled to their quadrupole mass spectrometer, voltages of the order of 600–700 V were applied to the capillary via the gold coating, and an electrostatically generated electrospray obtained. Wilm and Mann found that 1 µl of solvent tended to spray for approximately 40 min, giving flow rates of the order of 20–40 nl/min. According to their calculations, droplets emitted from the tip of the Taylor cone should have a diameter of about 200 nm, and at concentrations of about 1 pmol/µl this corresponds on average to one analyte molecule per droplet. The nano-ES interface is extremely efficient. Wilm and Mann [23] determined an analyte transfer efficiency of 1/390, compared with a value of 1/200,800 with conventional ES on the same instrument. Since their early publications on nano-ES, Wilm and Mann and co-workers [24–27] have generated an impressive series of papers detailing how nano-ES can be effectively used in protein structure analysis. The most important factors which make nano-ES so effective in mass spectrometry are the longevity and stability of the electrospray and the fact that the process is concentration-dependent. At flow rates of the order of 20 nl/min, 1 µl of sample will last for almost 1 h, which allows time for the optimisation of the electrospray and mass spectrometric conditions and also ample time for making numerous mass spectrometric measurements. In this manner, Mann, Wilm and co-workers have been able to analyse mixtures of peptides generated by digestion of proteins. The mixture is initially mass-mapped, then selected peptides are sequenced by tandem mass spectrometry, and the protein subsequently identified by a database search. In this manner, proteins separated on polyacrylamide gels have been identified at a subpicomole level.

5. Picospray mass spectrometry

Following on from Wilm and Mann's nano-ES, Valaskovic, McLafferty and co-workers [28] introduced "picospray", the name being chosen to reflect the picolitre sample volumes and picolitre/min flow rates attained. Pico-ES is similar to nano-ES, but lower flow rates are achieved by spraying from smaller i.d. capillaries. The gold-coated borosilicate capillaries of the Wilm and Mann type have an i.d. of 0.5–1 mm in the nontapered region. Valaskovic's picospray capillaries differ in that they are made of short lengths of fused silica (i.d. 5–20 µm) which are subsequently pulled to give a tapered tip which is then etched with hydrofluoric acid to give a 1–5 µm i.d. spraying orifice. Reduced flow rates are obtained with picospray tips (tube i.d. 5 µm, tip i.d. 2 µm, tip capacity 0.2 nl, flow rate 0.1–1.5 nl/min) as a consequence of the viscous flow within the small-bore capillaries. Valaskovic and co-workers [28] have obtained impressive data with pico-ES coupled to their FTICR instrument. For an 8.6-fmol loading and 150-amol consumption of cytochrome c (12359.3 Da), a mass accuracy of ± 0.1 Da was achieved at a resolution of 10^5. On their instrument, pico-ES was found to give an improvement in sensitivity of three orders of magnitude.

6. Nano- and micro-ES on magnetic sector instruments

A low flow-rate ES interface, with both nano-ES and micro-ES probes, has been installed on a high-resolution double-focusing magnetic sector mass spectrometer in our laboratory. Since installation more than 2 years ago, this has been the only ion source used with the instrument, and at no time have we considered it necessary to refit the previous, conventional ES interface. A wide range of samples have been successfully analysed and a number of novel compounds characterised. For example, we have been able to identify haemoglobin variants, and have sequenced a porcine pulmonary surfactant polypeptide [29]. A wide variety of solvents have been found to spray well, ranging from 100% chloroform to 99% water. In our experience, nano-ES gives not only a higher current than conventional ES (probably as a result of the more efficient sampling) but also, as it operates at a flow rate of 2–5 μl/h instead of 2–5 μl/min, an overall increase in sensitivity of at least two orders of magnitude. Micro-ES gives similar currents to nano-ES but at higher flow rates. However, micro-ES has the advantage that it can be coupled to capillary column-LC for the analysis of complex mixtures.

7. Conclusions

Since the early 1980s mass spectrometry has become an increasingly important analytical tool in biochemistry and biomedicine. With the introduction of MALDI and ES just over 10 years ago, mass spectrometry has proved to be effective in the characterisation of peptides and proteins. With the advent of micro-ES and nano-ES, mass spectrometry has now become the most sensitive method for the sequence analysis of peptides and for the identification of proteins via database searches. Micro-ES and nano-ES interfaces have been fitted to the complete range of mass spectrometers ranging from trapping instruments [21, 28, 30], scanning quadrupole instruments [18–20, 22–26], magnetic sector instruments [29, 31] and a range of hybrid instruments [27, 32]. Instrument manufacturers now provide extremely reliable and sensitive mass spectrometers, and perhaps the major challenge for the mass spectroscopist is no longer operation of the instrument but to be able to get the femtomole of sample into the instrument for subsequent analysis, as discussed in other chapters of this volume.

Acknowledgements
This work was supported by the Swedish Medical Research Council (project no. 03X-12551) and the Karolinska Institutet.

References

1 Thomson JJ (1913) *Rays of Positive Electricity and Their Applications to Chemical Analysis*. Longmans Green, London
2 Aston FW (1933) *Mass Spectra and Isotopes*. Edward Arnold, London
3 Barber M, Bordoli RS, Sedgwick RD, Tyler AN (1981) Fast atom bombardment of solids (FAB): a new ion source for mass spectrometry. *J Chem Soc Chem Commun* 325–327
4 Meng CK, Mann M, Fenn JB (1988) Electrospray ionization of some polypeptides and small proteins. In: *Proc 36th ASMS Conf on Mass Spectrometry and Allied Topics, San Francisco, CA, June 1988*. ASMS, East Lansing, MI, 711–772
5 Karas M, Hillenkamp F (1988) Laser desorption ionization of proteins with molecular masses exceeding 10,000 daltons. *Anal Chem* 60: 1299–2301
6 Dole M, Mach LL, Hines RL, Mobley RC, Ferguson LD, Alice MB (1968) Molecular beams of macroions. *J Chem Phys* 49: 2240–2247
7 Yamashita M, Fenn JB (1984) Electrospray ion source. Another variation on the free-jet theme. *J Phys Chem* 88: 4451–4459
8 Yamashita M, Fenn JB (1984) Negative ion production with the electrospray ion source. *J Phys Chem* 88: 4671–4675
9 Fenn JB, Mann M, Meng CK, Wong SF, Whitehouse CM (1989) Electrospray ionization for mass spectrometry of large biomolecules. *Science* 246: 64–71
10 Iribarne JV, Thomson BA (1976) On the evaporation of small ions from charged droplets. *J Chem Phys* 64: 2287–2294
11 Thomson BA, Iribarne JV (1979) Field induced ion evaporation from liquid surfaces at atmospheric pressure. *J Chem Phys* 71: 4451–4463
12 Kebarle P, Ho Y (1997) On the mechanism of electrospray mass spectrometry. In: RB Cole (ed): *Electrospray Ionization Mass Spectrometry: Fundamentals, Instrumentation and Applications*. John Wiley, New York, 3–63
13 Taylor GI (1964) Disintegration of water drops in an electric field. *Proc R Soc London Ser A* 280: 383–397
14 Pfeifer RJ, Hendricks CD (1968) Parametric studies of electrohydrodynamic spraying. *AIAA J* 6: 496–502
15 Fernández de la Mora J, Loscertales IG (1994) The current emitted by highly conducting Taylor cones. *J Fluid Mech* 260: 155–184
16 Tang L, Kebarle P (1991) Effect of the conductivity of the electrosprayed solution on the electrospray current. Factors determining analyte sensitivity in electrospray mass spectrometry. *Anal Chem* 63: 2709–2715
17 Tang L, Kebarle P (1993) Dependence of ion intensity in electrospray mass spectrometry on the concentration of the analyte in the electrosprayed solution. *Anal Chem* 65: 3654–3668
18 Gale DC, Smith RD (1993) Small volume and low flow-rate electrospray ionization mass spectrometry of aqueous samples. *Rapid Commun Mass Spectrom* 7: 1017–1021
19 Emmett MR, Caprioli RM (1994) Micro-electrospray mass spectrometry: ultra-high-sensitivity analysis of peptides and proteins. *J Am Soc Mass Spectrom* 5: 605–613
20 Andren PE, Emmett MR, Caprioli RM (1994) Micro-electrospray: zeptomole/attomole per microlitre sensitivity for peptides. *J Am Soc Mass Spectrom* 5: 867–869
21 Emmett MR, White FM, Hendrickson CL, Shi SD-H, Marshall AG (1998) *J Am Soc Mass Spectrom* 9: 333–340
22 Wilm MS, Mann M (1994) Electrospray and Taylor-cone theory: Dole's beam of macromolecules at last? *Int J Mass Spectrom Ion Processes* 136: 167–180
23 Wilm M, Mann M (1996) Analytical properties of the nanoelectrospray ion source. *Anal Chem* 68: 1–8
24 Wilm M, Neubauer G, Mann M (1996) Parent ion scan of unseparated peptide mixtures. *Anal Chem* 68: 527–533
25 Shevchenko A, Wilm M, Vorm O, Mann M (1996) Mass spectrometric sequencing of proteins from silver-stained polyacrylamide gels. *Anal Chem* 68: 850–858
26 Wilm M, Shevchenko A, Houthaeve T, Brelt S, Schweigerer L, Fotsis T, Mann M (1996) Femtomole sequencing of proteins from polyacrylamide gels by nano-electrospray mass spectrometry. *Nature* 379: 466–469

27 Shevchenko A, Chernushevich I, Ens W, Standing KG, Thomson B, Wilm M, Mann M
    (1997) Rapid *de novo* peptide sequencing by a combination of nanoelectrospray, isotopic
    labeling and a quadrupole/time-of-flight mass spectrometer. *Rapid Commun Mass Spectrom* 11: 1015–1024
28 Valaskovic GA, Kelleher NL, Little DP, Aaserud DJ, McLafferty FW (1995) Attomole-
    sensitivity electrospray source for large-molecule mass spectrometry. *Anal Chem* 67:
    3802–3805
29 Griffiths WJ, Gustaffson M, Yang Y, Curstedt T, Sjövall J, Johansson J (1998) Analysis of
    variant forms of porcine surfactant polypeptide-C by nano-electrospray mass spectrometry.
    Rapid Commun Mass Spectrom 12: 1104–1114
30 Figeys D, Ning Y, Aabersold R (1997) A microfabricated device for rapid protein identifi-
    cation by microelectrospray ion trap mass spectrometry. *Anal Chem* 69: 3153–3160
31 Veenstra TD, Tomlinson AJ, Benson L, Kumar R, Naylor S (1998) Low temperature aqueous
    electrospray ionization mass spectrometry of noncovalent complexes. *J Am Soc Mass Spectrom* 9: 580–584
32 Morris HR, Paxton T, Dell A, Langhorne J, Berg M, Bordoli RS, Hoyes J, Bateman RH
    (1996) High sensitivity collisionally-activated decomposition tandem mass spectrometry
    on a novel quadrupole/orthogonal-acceleration time-of-flight mass spectrometer. *Rapid Commun Mass Spectrom* 10: 889–896

Proteomics in Functional Genomics
ed. by P. Jollès and H. Jörnvall
© 2000 Birkhäuser Verlag Basel/Switzerland

MALDI-TOF mass spectrometry in protein chemistry

Peter Roepstorff

Department of Molecular Biology, Odense University, DK-5230 Odense M, Denmark

Summary. Mass spectrometry has in the last decade been accepted as a key analytical technique in protein chemistry. It is now the preferred technique for identification of proteins separated by one- or two-dimensional polyacrylamide gel electrophoresis, i.e. in proteome analysis. It is the dominating technique for determination of posttranslational modifications in proteins. The two ionization techniques presently widely used in protein studies are matrix-assisted laser desorption/ionization (MALDI) in combination with time-of-flight (TOF) mass analyzers and electrospray ionization (ESI) in combination with a variety of mass analyzers. In this chapter the principles and performance of MALDI-TOF mass spectrometry will be described as well as the application of this technique to a variety of applications.

Introduction

The rapidly expanding use of gene technology in the last 2 decades has dramatically changed the focus of protein chemistry. The sequence of the human genome is expected to be concluded soon after the turn of the century, and sequencing of genomes from a number of other higher organisms, including mammals and plants, it also progressing rapidly. A consequence of this development is that the importance of *de novo* protein sequencing has been dramatically reduced, and the focus has changed to proteomics, i.e. identification of the proteins expressed in a given cell type at a given time relative to the known genomic or complementary DNA (cDNA) sequences [1], and to characterization of the proteins in terms of primary structure, including transient modifications. In addition, new rapid and sensitive methodologies must be developed to study protein higher-order structure and protein interactions. It must be realized, however, that genome sequences will still be unknown for a majority of species. For some, expressed sequence tag data (ESTs) will be available, for others cDNA libraries, and for most, no information at all. Therefore, there is a need for developing methods that allow rapid generation of sufficient sequence information for identification of the corresponding EST or cDNA and subsequent cloning. Moreover, development of new concepts that would allow correlation between the proteins from any species with those from species with known genomes would be highly desirable.

Concurrent with the expansion of gene technology, mass spectrometry (MS) has undergone an equally rapid development. Two new ionization techniques, matrix assisted laser desorption ionization (MALDI) [2] and electrospray ionization (ESI) [3] have opened the field of analysis of large biomolecules. Both techniques allow analysis of proteins in excess of 200 kDa, in addition to complex peptide mixtures. These mass spectrometric techniques ideally complement the information derived by genetic techniques. The precise mass of any given peptide or protein is a highly specific characteristic of the compound which can be matched with calculated masses obtained from sequence information available in databases. The two mass spectrometric techniques are highly complementary, but MALDI-MS is perhaps the more sensitive of the two. Sample preparation is simple and rapid, and a relatively high level of contaminating compounds can be tolerated. Alternatively, ESI can be coupled to high performance liquid chromatography (HPLC) and/or capillary electrophoresis (CE) for on-line separation and mass analysis. ESI-MS allows precise molecular mass determination of proteins, and when combined with tandem mass spectrometry it is possible to generate sequence information from peptides and small proteins. This can be achieved with either triple quadrupole or trapping mass analyzers. In our laboratory we have access to MALDI-time-of-flight (TOF), ESI-triple quadrupole and ESI-ion trap instruments. MALDI-TOF mass spectrometry, however, is nearly always the first technique chosen in any protein study and in many cases the only technique required. In this chapter, the principles and present performance of MALDI-TOF will be described. In addition, examples of applications of MALDI-TOF from our current studies on protein identification and characterization of secondary modifications will be given. Future perspectives for studies of protein higher-order structure and protein interactions will be outlined.

1. Principles and performance of the MALDI-TOF instrument

1.1. The instrument, principle, resolution and mass accuracy

The principle of a contemporary MALDI-TOF instrument is illustrated in Figure 1. A solution of the analyte (a protein, peptide or mixture) is mixed on the mass spectrometric target with a solution of an appropriate matrix. After cocrystallization, the matrix/analyte layer is irradiated with a short laser pulse, resulting in desorption of matrix and analyte molecules and ions. The ions are accelerated into the TOF mass analyzer, which consists of a field-free flight tube and travels in the linear operation mode to detector 1 at the end of the instrument. Once the flight time (T) is known, then the mass/charge ratio (m/z) of the analyte ions can be calculated based on the equation

$$T = C_1 \sqrt{m/z} + C_2 \qquad (1)$$

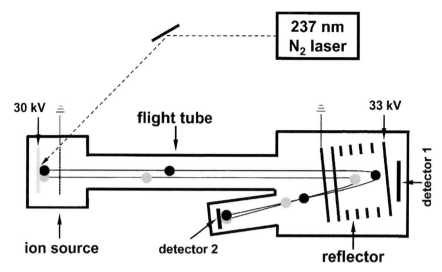

Figure 1. Principle of a reflector MALDI-TOF instrument.

where C_1 and C_2 are instrumental constants which can be determined with compounds of known mass.

The linear TOF spectrum is limited in resolution, and thus mass accuracy, because an initial energy spread of the ions will cause a spread in the flight time of the ions. This energy spread can be partially compensated by reflecting the ions in an electrostatic mirror and detecting the ions with detector 2 (RF-TOF). In addition, it was recently discovered that the energy spread of the ions could be reduced by applying the acceleration voltage with a slight delay relative to the laser pulse, termed delayed extraction (DE). The resolution in linear and reflector mode [4] is thereby increased. With modern MALDI-TOF instruments equipped with delayed extraction, resolution greater than 2000 can be obtained in linear mode and greater than 10,000 in reflector mode. Such resolution enables full isotopic resolution for molecules up to 15 kDa in reflector mode. The ^{12}C-only ion, however, will be of negligible intensity for molecules beyond approximately 5 kDa and therefore difficult to assign. In addition, all ions undergo a certain degree of decomposition after acceleration, termed post-source decay (PSD). This decomposition causes an energy spread which degrades the resolution. Moreover, ions decomposing in the first field-free region will not be detected in reflector mode because their kinetic energy deficiency cannot be compensated by the electrostatic mirror. This will result in decreased sensitivity because the degraded fraction of the desorbed ions will not be detected. In linear mode the decomposed ions will be detected at the same flight time as the molecular ions but with a slightly higher energy spread, resulting in loss of resolution. The tendency to fragment increases with increasing size of the analyte molecules. Therefore, it is

Table 1. Mass accuracy and sensitivity routinely achieved by MALDI-MS of peptides and proteins

Mass range		0.5–5 kDa	5–20 kDa	> 20 kDa
Mass accuracy	DE-RF-TOF	30 ppm	50–100 ppm	–
	DE-linear-TOF	100 ppm	100–200 ppm	0.02–0.1%
Sensitivity	all modes	0.1–1 fmol	1–10 fmol	0.1–1 pmol

generally advantageous to record the spectra in linear mode and determine the isotopically averaged mass for molecules beyond 5–10 kDa. Consequently, the mass accuracy achieved will depend on the type of analyte, its molecular size and the recording mode selected (Tab. 1).

1.2. The matrix

The requirements of the matrix molecules are that they must have ultraviolet (UV) adsorption at the wavelength of the laser (typically a 237-nm N_2 laser), low volatility and the ability to transfer protons to the analyte molecules. Typical matrices suitable for MALDI-MS of proteins include a variety of cinnamic acid and hydroxylated benzoic acid derivatives. The general functions of the matrix are to provide a medium which can adsorb the energy of the laser pulse, thereby causing desorption of the analyte molecules in an expanding plume, to ionize the desorbed analyte molecules and to prevent aggregation of the analyte molecules.

1.3. Sensitivity

Sample preparation prior to mass spectrometric analysis and preparation of the matrix/analyte layer are crucial for the sensitivity which can be obtained by MALDI-MS analysis and must be adapted to the specific analyte based on its type, size and previous history. Thus, different matrices and preparation methods are optimal for analysis of peptide mixtures, large proteins or glycoproteins [5]. The major limiting factor for sensitivity is chemical noise caused by impurities in the sample, by the matrix and by a number of unknown decomposition and/or association reactions occurring during the desorption process. Only the first two limitations have been successfully addressed. Sensitivity in the low attomole range has been achieved for peptides by preparation of thin matrix layers followed by application of the analyte such that it is incorporated in the outer matrix layer. This method allows removal of salts present in the analyte solution by a simple washing procedure [6]. Low attomole sensitivity has also been achieved by generating very small matrix spots using nanoliter volumes of matrix and analyte solution [7]. Recently, similar sensitivities have been obtained with a combined prepurification, concentration and sample application procedure. The method is based on adsorption of the analyte on a nanoliter bed volume reversed-phase column prepared in an Eppendorf GeLoader tip. The column is washed and the analyte eluted with a few nanoliters of matrix solution onto the mass spectrometric target [8]. The sensitivity is reduced with increasing molecular weight, and for analysis of proteins it is generally two to three orders of magnitude less than for peptides, i.e. in the femtomole range (Tab. 1).

1.4. Structural information

Structural information can be readily generated by mass spectrometric mapping of the mixture derived by digestion of the protein with specific endoproteases, e.g. trypsin and endoproteinases AspN or GluC. MALDI-MS is the foremost mass spectrometric technique for direct analysis of the resultant peptide mixtures. The signal intensity, however, does not necessarily reflect the quantities of the different peptides in the mixture. Some peptide signals may be entirely suppressed, presumably due to competition for charge or for optimal positions in the matrix. Combining information obtained with different matrices [9], from spectra recorded in positive and negative ion mode, or from digestion with different enzymes [10] often results in complete coverage of the sequence.

Sequence information on peptides can be generated in MALDI-MS by taking advantage of the PSD mentioned above. By a stepwise decrease of the reflector voltages it is possible to focus fragment ions with different m/z ranges on the detector and generate a PSD spectrum [11]. Since only a small fraction of the generated molecular ions undergo PSD, substantially larger amounts of analyte are required to obtain PSD spectra comparable to standard peptide spectra. In addition, the fragmentation process cannot be controlled, and the proportion of PSD and sites of fragmentation vary considerably between peptides. Only rarely is it possible to obtain full sequence information. To generate more controlled fragmentation, several instrument manufacturers now include a collision cell in the flight tube to perform collision-induced dissociation. In instruments epuipped with delayed extraction [4] it has recently been observed that upon MALDI analysis of proteins, a considerable quantity of fragmentation occurs in the ion source prior to initiation of the acceleration voltage, termed in-source decay. Such fragment ions may yield long regions of sequence-specific ions [12, 13].

An alternative to generation of sequence-specific fragment ions in the mass spectrometer is to generate peptide mixtures containing "sequence ladders". Thus C-terminal sequence ladders can be obtained by digestion with carboxypeptidases [14]. N-terminal ladders are generated by Edman degradation using a low percentage of phenylthiocarbamate rather than phenyoisothiocarbamate in the coupling reaction [15]. MALDI mass spectra of mixtures generated by mild acid hydrolysis of peptides [16–18] and proteins [19] often contain a substantial proportion of sequence information.

2. Applications

2.1. Proteome analysis and identification of proteins by MALDI mapping

Proteome analysis involves two essential steps. Firstly, separation and visualization of the proteins and, secondly, identification of the proteins relative to the genomic sequence, if known. Two-dimensional polyacrylamide gel electrophoresis (2D-PAGE) [20–22] is the only technique presently available for separation of all, or the majority of, the proteins from a given cell type. Identification of the proteins present in a spot on a 2D gel has traditionally been performed by micro-Edman degradation [23]. Due to the limited sensitivity of this technique, however, the demand for an alternative method was obvious. Proteolytic digestion of a given protein with a sequence-specific protease, e.g. trypsin, which cleaves at Lys and Arg residues, produces a set of peptides which serve as a unique fingerprint. In 1993, a number of groups independently demonstrated that peptide maps produced by mass spectrometry (MS) allowed identification of proteins relative to full-length protein sequence information (derived from genome, cDNA or protein sequencing) [24–28]. Comparison of the experimentally measured set of peptide masses against those predicted for each entry in a sequence database will retrieve a number of matching protein sequences. This list of sequences is then evaluated and scored, e.g. by considering the origin of species, the calculated protein mass and pI, the experimental peptide mass error, the number of peptide mass matches and the amino acid sequence coverage of the assigned peptides. As an alternative, peptide sequence analysis by electrospray ionization tandem mass spectrometry (ESI-MS/MS) can also be utilized in protein identification. Sequence databases are queried by either uninterpreted tandem mass spectra using a pattern recognition scheme [29] or by partially interpreted spectra using the peptide sequence tag concept. This allows identification of proteins based on the molecular mass and a partial sequence from a single proteolytically derived peptide [30]. Recently, peptide sequence tags have also been successfully derived by PSD of selected ions in the peptide maps derived by MALDI-MS [31]. Since then numerous groups have further developed these concepts and applied them to identification of proteins isolated by 1D- or 2D-PAGE [32–35]. Peptide mapping by MALDI-MS is the simplest and most sensitive of these techniques, and is gradually reaching a confidence level comparable to sequence-based identification. This is primarily due to improved mass accuracy and sample preparation methods and

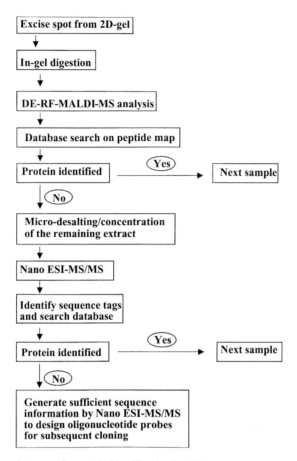

Figure 2. General strategy for protein identification by MS.

availability of complete genomic information for a number of organisms. It is therefore the method of choice in our general strategy for identification of gel-separated proteins (Fig. 2).

The identification by MALDI mapping of a protein present in a spot on a 2D gel can be exemplified by the identification of a protein in a study of yeast proteins. The yeast proteins were separated by 2D-PAGE, and the protein spot excised and digested *in situ* following our standard protocol [32]. Ten percent of the extracted peptide mixture was analyzed by DE-RF-MALDI-TOF mass spectrometry, resulting in the mass spectrum shown in Figure 3a. Each ion signal in the spectrum corresponds to a protonated peptide molecular ion. A comprehensive protein sequence database was queried with this set of 26 tryptic peptide masses entered in the peptide search program [25, 30]. Yeast GBLP was retrieved as the highest scoring sequence with 16 peptide mass matches (Fig. 3b). The next protein

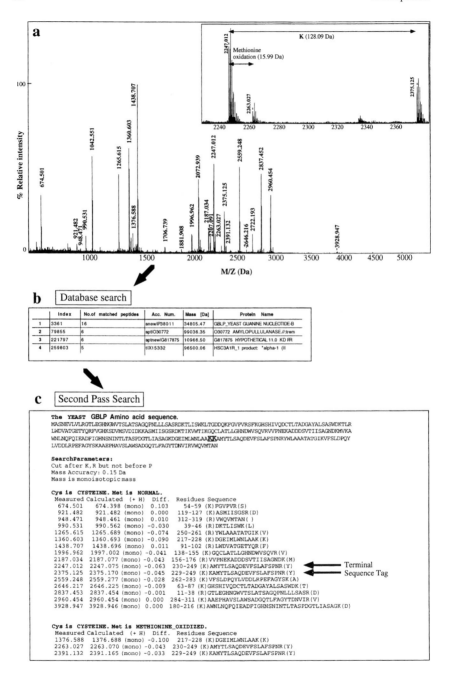

Figure 3. (a) MALDI mass spectrum generated from a protein isolated by 2D-PAGE of a yeast extract and digested *in situ*. (b) Result of a database search using the Peptide Search program. (c) Result of the second pass search. Peptide pairs resulting from cleavage at double basic residues are labelled, and the corresponding peaks indicated in the inset (a).

sequences in the list had six mass matches, or less. Using the second pass search facility of the program, detailed evaluation of the data revealed that three additional peptide masses could be assigned to the sequence when assuming oxidation of methionine to methionine sulphoxide (mass increment of 16 Da). Further confirmation of the assigned sequence was obtained by examination of the spectrum for alternative tryptic cleavages at double basic residues (Fig. 3a, c). The combination of the mass accuracy obtained with DE-RF-MALDI-MS and examination of the spectra for peptide pairs derived from partial oxidation of Met residues and alternative cleavage at adjacent basic residues improves the fidelity of the identification to certainty.

Peptide mapping by MALDI-MS normally allows identification of all the proteins when studying organisms for which the complete genome sequence is available and a large proportion of the proteins from other organisms. If identification is not achieved by MALDI mass mapping, generation of partial sequence information (step 2 of the strategy in Fig. 2) often allows identification in cases when only partial protein sequence information is available in the databases, e.g. as ESTs [37] or based on homology to protein sequences from other organisms. If the protein after this step is classified as unknown, two alternatives are available. Sufficient sequence information for construction of oligonucleotide probes and subsequent cloning can be generated by MS/MS of the peptides in the mixture extracted from the gel [38, 39], or if sufficient protein can be obtained, it can be isolated and analyzed for sequence. Although it has often been claimed that it is possible to sequence complete proteins using only mass spectrometry, it is our experience that our previously published strategy [40], which is based on a combination of MS and Edman degradation, has proven to be the most successful approach.

2.2. MS for the determination of secondary protein modifications

As reported in many recent reviews and monographs [e.g. 41, 42], MS has been the preferred method for characterization of secondary modifications in natural and recombinantly produced proteins. General strategies for characterization of secondary modifications and verification of engineered proteins are described in [41]. The first step is to measure the molecular mass of the intact protein by MALDI- or ESI-MS. When performed with sufficient accuracy, this single measurement allows the determination of the degree of modification of the protein, provided that its sequence is known. The next step is to generate site-specific information by direct mass spectrometric peptide mapping of a mixture derived by proteolytic cleavage of the protein by MALDI-MS, or following separation of the components by liquid chromatography (LC) combined off-line with MALDI-MS or on-line with ESI-MS and/or MS/MS. In our laboratory all four

approaches have been used to characterize recombinant and natural proteins [43–47]. Since direct peptide mapping is rapid and sensitive, this is
always our first approach. Moreover, it is also the first step in the strategy
for identification of proteins in gels and examination of the peptide maps
for the presence of posttranslationally modified peptides and it is a natural
next step after identification of the protein. When the molecular mass of the
protein can be measured, the total mass of the modifying group is also
known, and an incomplete peptide map will often be sufficient to locate the
modified sites and the nature of the modifications. Unfortunately, this
information is normally not available for gel-separated proteins, although
it can be obtained under favorable conditions either after blotting to a
PVDF membrane [48] or after electroelution [49–51]. Consequently, complete or nearly complete sequence coverage is essential to determine all
modifications present in a gel-separated protein. The sequence coverage
obtained from mapping gel-separated proteins varies between 30 and nearly 100% depending on the characteristics of the protein and the quantity of
protein in the spot. This is generally insufficient to ensure that all modified
sites are located.

To illustrate the type of information contained in a MALDI peptide map
of a protein in a gel spot, let us reconsider the example shown in Figure 3.
Examination of the second pass search (Fig. 3c) shows that the sequence
coverage is 84%. One of the missing peptides is the N-terminal tryptic peptide 1–10. This peptide will not be observed in the search if the sequence
in the database is derived from genome and cDNA sequencing, because
the N-terminal Met residue is not part of the actual protein sequence.
Examination of the spectrum does not reveal the presence of a peak corresponding to the peptide 2–10 (MH$^+$ expected at m/z 1000.57). Instead, an
intense unassigned peak at m/z 1042.55 can be tentatively assigned to the
N-acetylated N-terminal peptide (calculated m/z 1042.59). The identity of
this acetylated peptide can be confirmed by PSD or MS/MS of the ion, or
by a simple derivatization reaction on an aliquot of the remaining extract.
Inclusion of the N-terminal peptide increases the sequence coverage to
87%. The remaining missing peptides are one 11-residue peptide, a heptapeptide, two pentapeptides and four small di- to tripeptides with molecular
masses below 600 Da. These small peptides are of marginal value for protein identification and are normally not observed. To avoid saturation of the
detector by matrix ions [52], signals below m/z 600 are not detected.
Examination of the sequence of the remaining four peptides shows that
three of them may be modified either by phosphorylation or O-glycosylation. It is probable that the heptapeptide with the sequence LTGDDQK
is phosphorylated because it contains a consensus sequence for casein
kinase II phosphorylation (underlined). No peaks were present in the spectrum which are indicative of peptides with phosphorylation, or with man-
nosyl type O-glycosylation (the most probable in yeast). So from the present spectrum it is not possible to determine whether these peptides are

modified, because the signals are suppressed. If a peak is present which might be assigned to a phosphorylated peptide, the phosphorylation can be confirmed by treatment of an aliquot of the extract with alkaline phosphatase. A decrease in mass by 80 Da (or multiples thereof) is observed for the signals corresponding to phosphorylated peptides [53]. If the presence of phosphorylated peptides is suspected but corresponding signals are absent, as in this example, it is possible to test further for the presence of phosphorylated peptides by attempts at peptide isolation by affinity chromatography on a small iron-containing column followed by identification of the bound peptides by MALDI-MS [54]. The reduced number of peptides in the resultant mixture decreases the likelihood of signal suppression.

MALDI-MS is also ideally suited for analysis of glycoproteins [55]. Two types of glycosylation exist: N-linked glycans attached to Asn residues and O-linked glycans attached to Ser or Thr residues. Heterogeneous branched sugar chains are typically composed of hexosamines, hexoses and sialic acids. The extensive heterogeneity and nonlinear structure of glycans make structural elucidation a challenge by any method. Mass spectrometric analyses are typically performed on isolated glycopeptides because the glycopeptides exhibit less signal intensity than unmodified peptides and therefore are difficult to observe by direct peptide mapping. The combination of digestion with sequence-specific glycolytic enzymes and analysis by MS is a popular approach for investigation of carbohydate structures in proteins. The elucidation of a triantennary complex-type N-linked glycan structure in the major cat allergen [46] is illustrated in Figure 4. This strategy has only in very few cases been applied to glycoproteins in gels. In an example from our laboratory, several interferon-γ glycoforms were separated by SDS/PAGE [56]. Following electroelution from the gel, intact mass determination by MALDI-MS of three protein isoforms suggested that the proteins contained 0, 1 or 2 glycosylated sites, respectively. In-gel digestion followed by MALDI peptide mass mapping and sequential exoglycosidase digestion identified one glycosylated peptide and its glycan structure. The second glycosylated peptide, however, was not detected by MALDI-MS analysis of the peptide mixture. Microbore HPLC was utilized for its isolation followed by structural characterization by MALDI-MS and glycosidase treatment. It was estabished that the complex spectra obtained by MALDI-MS of these glycopeptides reflected glycosylation heterogeneity rather than decomposition and fragmentation of glycans during mass spectrometric analysis [57]. The relative intensity of the peaks representing the different glycoforms is indicative of the quantitative relationship between the different forms. Therefore, sequential glycosidase digestion can be avoided in cases where it is of interest to characterize the site-specific carbohydrate mass profile of a glycoprotein, rather than detailed structural characterization of the glycans [47, 58, 59]. This is known as glycoprofiling. The site specificity in the above-described strategy is obtained by proteolytic cleavage between the glycosylated sites in the protein. This

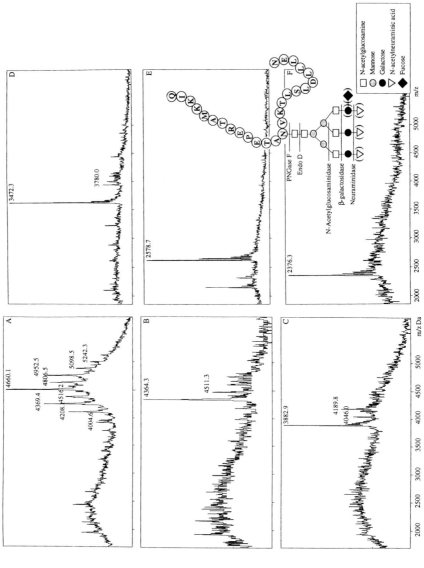

Figure 4. Analysis by MALDI-MS of a glycopeptide from the major cat allergen Fel d1. MALDI mass spectra of the glycopeptide (A) and after sequential treatment with neuraminidase (B), β-galactosidase (C), N-acetylglucosamidase (D), endoglucanase D (E) and PNGase F (F). The structures determined by MS are given with the sugar residues responsible for the heterogeneity indicated in parentheses and the specificity of the enzymes indicated by the lines. Adapted from [46].

can be difficult to achieve in proteins with densely O-glycosylated regions, e.g. in mucin-type O-glycosylated peptides. PSD analysis may allow site-specific assignment, but O-glycosyl bonds fragment more readily than peptide bonds, with the result that loss of the glycans dominates the spectra, and sequence ions with intact glycans are rare and of very low abundance. Assignment of the glycosylated sites is therefore difficult [60]. Recently it was demonstrated that mild acid gas-phase hydrolysis can be optimized to hydrolyze peptide bonds rather than the O-glycosyl bonds with preference for bonds N-terminal to Ser and Thr and C-terminal to Asp. MALDI-MS of the resultant mixtures allows assignment of the O-glyco-sylated sites in densely glycosylated regions [18].

3. Conclusions and future perspectives

MALDI-MS emerged as a new technology 10 years ago, and commercial instruments have only been available for approximately 8 years. In spite of this short period, it is already a tool of major importance in protein chemistry. The utility of MALDI-MS for site-specific assignment of sec-ondary protein modifications is well established. Although based on very recent achievements, the role of MALDI-MS for identification of proteins isolated by gel electrophoresis or otherwise is rapidly being recognized. The major reasons for the quick acceptance of this technique are the simplicity, user-friendliness and affordability of the instrumentation as well as the spectacular recent progress that has been achieved in sensitivity and mass accuracy.

The next important area for application of MALDI-MS in protein che-mistry undoubtedly will be in studies of protein higher-order structure and protein interactions. MS is gradually entering this field. Most applications have been based on analysis by ESI-MS either to monitor deuterium exchange in the liquid phase or to measure non-covalent complexes [61, 62]. Although noncovalent protein complexes have been observed by MALDI-MS [63], it is not considered to be suitable for such studies because the complexes normally dissociate upon mixing with matrix solu-tion. Structural information on native proteins can be obtained by chemi-cally labeling surface-accessible residues in the native form of the protein [64] or by limited proteolysis [65] followed by peptide mapping by MAL-DI-MS. Similarly, residues shielded when the protein is involved in an interaction can be mapped [66, 67], or the interacting regions can be cross-linked and the positions of the cross-links determined by MALDI mass mapping [68]. MALDI-MS has also been applied for interaction studies in combination with affinity-based techniques. Thus, incubation of a protein digest with antibodies was used to isolate antibody-binding epitopes fol-lowed by precipitation of the antibody/antigen complex with agarose beads. The beads were then washed and subjected to MALDI-MS analysis

[69]. A similar approach using the avidin/biotin interaction has been used to isolate biotinylated peptides [70] and to identify a biotin-labeled cross-link between a receptor and its ligand [71]. Streptavidin-coated magnetic beads were used to isolate DNA-binding proteins by first binding biotinylated double-stranded DNA to the beads, followed by incubation with nuclear cell extracts [67]. In an alternative approach, antibodies immobilized directly on the mass spectrometric target were used to isolate the corresponding antigen from protein extracts or serum [72]. This concept was further extended to the direct analysis by MALDI-MS of a ligand bound to antibodies immobilized on a chip. These interaction studies were accomplished by surface plasmon resonance using a BiaCore instrument [73]. As an alternative, the ligand bound to the Bia-chip can be eluted and analyzed by MALDI or ESI-MS [74]. There is no doubt that these examples represent the beginning of a number of new applications, and that MALDI-MS will play an important future role in studies of protein higher-order structure and protein interactions.

Acknowledgments
Ekatarina Mirgorodskaya, Keyrin Bennett and Martin R. Larsen are acknowledged for help in preparing the figures and for proofreading the manuscript. The Danish Biotechnology Program and the Danish National Research Foundation are acknowledged for financial support.

References

1 Wilkins MR, Pasquali C, Appel RD, Ou K, Golaz O, Sanchez J-C, Jan JX, Gouley AA, Humphrey-Smith I, Williams KL et al (1996) From proteins to proteomers: large-scale protein identification by two-dimensional electrophoresis and amino acid analysis. *Bio/Technology* 14: 61–65

2 Karas M, Hillenkamp F (1988) Laser desorption ionization of proteins with molecular masses exceeding 10,000 daltons. *Anal Chem* 60: 2299–2301

3 Fenn JB, Mann M, Meng CK, Wong SF, Whitehouse CM (1989) Electrospray ionization for the mass spectrometry of large biomolecules. *Science* 246: 64–71

4 Vestal ML, Juhasz P, Martin SA (1995) Delayedextraction matrix-assisted laser desorption/ionization time-of-flight mass spectrometry. *Rapid Comm Mass Spectrom* 9: 1044–1050

5 Kussmann M, Nordhoff E, Nielsen HR, Larsen M, Haebel S, Mirgorodskaya E, Jensen C, Roepstorff P (1997) Sample preparation in MALDI-MS designed for various peptide and protein analytes. *J Mass Spectrom* 32: 593–601

6 Vorm O, Roepstorff P, Mann M (1994) Improved resolution and very high sensitivity in MALDI-TOF of matrix surfaces made by fast evaporation. *Anal Chem* 66: 3281–3287

7 Jespersen S, Niessen WM, Tjaden UR, Greef J vd, Litborn E, Lindberg U, Roeraade J (1994) Attomole detection of proteins by matrix assisted lasr desorption/ionization mass spectrometry with the use of picoliter vials. *Rapid Comm. Mass Spectrom* 8: 581–584

8 Gobom J, Nordhoff E, Mirgorodskaya E, Ekman R, Roepstorff P (1999) A sample purification and preparation technique based on nano-scale RP-columns for the sensitive analysis of complex peptide mixtures by MALDI-MS. *J Mass Spectrom* 34: 105–116

9 Cohen S, Chait BT (1996) Influence of matrix solution conditions on the MALDI-MS analysis of peptides and proteins. *Anal Chem* 68: 31–37

10 Kussmann M, Lässing U, Stürmer CAO, Przybylski M, Roepstorff P (1997) MALDI mass spectrometric peptide mapping of hte neural cell adhesion protein neurolin purified by sodium dodecyl sulfate polyacrylamide gel electrophoresis or acidic precipitation. *J Mass Spectrom* 32: 483–493

11 Spengler B, Kirsch D, Kaufmann R, Jaeger E (1992) Peptide sequencing by matrix assisted laser desorption/ionization mass spectrometry. *Rapid Commun Mass Spectrom* 6: 105–108

12 Reiber DC, Grower TA, Brown RS (1998) Identifying proteins and matrix-assisted laser desorption/ionization in source fragmentation data combined with databasesearching. *Anal Chem* 70: 673–683

13 Reiber DC, Brown RS, Weinberger S, Kenny J, Bailey JC (1998) Unknown peptide sequencing using matrix assisted laser desorption/ionization and in-source decay. *Anal Chem* 70: 1214–1222

14 Patterson DH, Tarr GE, Regnier FE, Martin SA (1995) C-terminal ladder sequencing via matrix-assisted laser desorption/ionization mass spectrometry coupled with carboxypeptidase Y time-dependent and concentration-dependent digestions. *Anal Chem* 67: 3971–3978

15 Chait BT, Wang R, Beavis RC, Kent SBH (1993) Protein ladder sequencing. *Science* 262: 89–92

16 Vorm O, Roepstorff P (1994) Peptide sequence information derived by partial acid hydrolysis and matrix assisted laser desorption ionization mass spectrometry. *Biol Mass Spectrom* 23: 734–740

17 Zubarev RA, Chivanov VD, Håkansson P, Sundqvist BUR (1994) Peptide sequencing by partial acid hydrolysis and high resolution plasma desorption mass spectrometry. *Rapid Commun Mass Spectrom* 8: 906–912

18 Mirgorodskaya E, Hassan H, Wandall HH, Clausen H, Roepstorff P (1999) Partial vapor phase hydrolysis of peptide bonds: an effective method for mass spectrometric determination of O-glycosylated sites in glycopeptides. *Anal Biochem* 269: 54–65

19 Gobom J, Mirgorodskaya E, Nordhoff E, Roepstorff P (1998) Mass spectrometric peptide mapping using a new chemical method for sequence specific cleavage of proteins. Proc. 46th ASMS Conference of Mass Spectrometry and Allied Topics, Orlando, Florida, 30 May–4 June

20 O'Farrell PH (1975) High resolution two-dimensional electrophoresis of proteins. *J Biol Chem* 250: 4007–4021

21 Klose J (1975) Protein mapping by combined isoelectric focusing and electrophoresis of mouse tissues. A novel approach to testing for induced point mutations in mammals. *Humangenetik* 26: 231

22 Fey SJ, Nawrocki A, Larsen MR, Görg A, Roepstorff P, Skews GN, Williams R, Larsen PM (1997) Proteome analysis of *Saccharomyces cerevisiae*: A methodological outline. *Electrophoresis* 18: 1361–1372

23 Celis JE, Ratz GP, Madsen P, Gesser B, Lauridsen JB, Kwee S, Rasmussen HH, Nielsen HV, Cruger D, Basse B et al (1989) Comprehensive, human cellular protein databases and their implication for the study of genome organization and function. *FEBS Lett* 244: 247–254

24 Henzel WJ, Billeci TM, Stults JT, Wong SC (1993) Identifying proteins from two-dimensional gels by molecular mass searching of peptide fragments in protein sequence databases. *Proc Natl Acad Sci USA* 90: 5011–5015

25 Mann M, Højrup P, Roepstorff P (1993) Use of mass spectrometric molecular weight information to identify proteins in sequence databases. *Biol Mass Spectrom* 22: 338–345

26 Pappin DJC Højrup P, Bleasby AJ (1993) Protein identification by peptide mass fingerprinting. *Curr Biol* 3: 327–332

27 James P, Quadroni M, Carafoli E, Gonnet G (1993) Protein identification by mass profile fingerprinting, *Biochem Biophys Res Commun* 195: 58–64

28 Yates JR, Speicher S, Griffin PR, Hunkapiller T (1993) Peptide mass maps: a highly informative approach to protein identification. *Anal Biochem* 214: 397–408

29 Engl JK, McCormack AL, Yates JR (1994) An approach to correlate tandem mass spectra of peptides with amino acid sequences in protein databases. *J Am Soc Mass Spectrom* 5: 976–989

30 Mann M, Wilm MS (1994) Error tolerant identification of peptides in sequence databases by peptide sequence tags. *Anal Chem* 66: 4390–4399

31 Gevaert K, Demol H, Skylyarova T, Vandekerckhove J, Houthave T (1998) A peptide concentration and purification method for protein characterization in the subpicomole range using matrix assisted laser desorption/ionization-post source decay (MALDI-PSD) sequencing. *Electrophoresis* 19: 909–917

32 Patterson SD, Aebersold R (1995) Mass spectrometric approaches for the identification of gel-separated proteins. *Electrophoresis* 16: 1791–1814

33 Shevchenko A, Jensen ON, Podtelejnikov AV, Sagliocco F, Wilm M, Vorm O, Mortensen P,
   Shevchenko A, Boucherie H, Mann M (1996) Linking genome and proteome by mass spectrometry: large scale identification of yeast proteins from two dimensional gels. *Proc Natl
   Acad Sci USA* 93: 14440–14445
34 Dongre A, Eng J, Yates JR (1997) Emerging tandem-mass-spectrometry techniques for the
   rapid identification of proteins. *Trends Biotechnol* 15: 418–425
35 Patterson SD (ed) (1998) Paper symposium. Mass spectrometry in electrophoresis. *Electrophoresis* 19: 883–1054
36 Fey SJ, Nawrocki A, Larsen MR, Görg A, Roepstorff P, Skews GN, Williams R, Mose
   Larsen P (1997) Proteome analysis of *Saccharomyces cerevisiae*: a methodological outline.
   Electrophoresis 18: 1361–1372
37 Mann M (1996) A shortcut to interesting human genes: peptide sequence tags, ESTs and
   computers. *Trends Biol Sci* 21: 494–495
38 Wilm M, Shevchenko A, Houthaeve T, Breit S, Schweigerer L, Fotsis T, Mann M (1996)
   Femtomole sequencing of proteins from polyacrylamide gels by nanoelectrospray mass
   spectrometry. *Nature* 379: 466–469
39 Shevchenko A, Wilm M, Mann M (1997) Peptide sequencing by mass spectrometry for
   homology searches and cloning of genes. *J Protein Chem* 16: 481–490
40 Roepstorff P, Højrup P (1993) A general strategy for the use of mass spectrometric molecular weight information in protein purification and sequence determination. In: *Methods in
   Protein Sequence Analysis*: K Imahori, F Sakiyama (eds) Plenum Press, New York, 149–156
41 Andersen JS, Svensson B, Roepstorff P (1996) Electrospray ionization and matrix-assisted
   laser desorption/ionization mass spectrometry: powerful analytical tools in recombinant
   protein chemistry. *Nature Biotechnol* 14: 449–457
42 Burlingame AL, Carr SA (eds) (1996) *Mass spectrometry in biological sciences*. Humana
   Press, Totowa, NY
43 Andersen JS, Søgaard M, Svensson B, Roepstorff P (1994) Localization of an O-glycosylated site in recombinant barley α-amylase 1 produced in yeast and correction of the
   amino acid sequence using matrix assisted laser desorption ionization. *Biol Mass Spectrom*
   23: 547–554
44 Brody S, Andersen JS, Kannagara CG, Melgaard M, Roepstorff P, von Wettstein D (1995)
   Characterization of different spectral forms of glutamate 1-semialdehyde amino transferase
   by mass spectrometry. *Biochemistry* 34: 15918–15924
45 Juge N, Andersen JS, Tull D, Roepstorff P, Svensson B (1996) Overexpression, purification,
   and characterization of recombinant barley α-amylases 1 and 2 secreted by the methylo-
   trophic yest *Pichia pastoris*. *Protein Expression and Purification* 8: 204–214
46 Kristensen AK, Schou C, Roepstorff P (1997) Determination of isoforms, N-linked glycan
   structure and disulfide bond linkages of the major cat allergen Fel d 1 by a mass spectrometric approach. *Biol Chem* 378: 899–908
47 Rahbek-Nielsen H, Roepstorff P, Reischl H, Wozny M, Koll H, Haselbeck A (1997) Glycopeptide profiling of human urinary erythropoietin by matrix-assisted laser desorption
   ionization mass spectrometry. *J Mass Spectrom* 32: 948–958
48 Strupat K, Karas K, Hillenkamp F, Eckerskorn C, Lottspeich F (1994) Matrix assisted laser
   desorption/ionization mass spectrometry of proteins electroblotted after polyacrylamide gel
   electrophoresis. *Anal Chem* 66: 464–470
49 Haebel S, Jensen C, Andersen SO, Roepstorff P (1995) Isoforms of a cuticular protein from
   larvae of the meal beetle, *Tenebrio molitor*, studied by mass spectrometry in combination
   with Edman degradation and 2D-PAGE. *Protein Sci* 4: 394–404
50 Jensen C, Haebel S, Andersen SO, Roepstorff P (1997) Towards monitoring of protein
   purification by matrix-assisted laser desorption ionization mass spectrometry. *Int J Mass
   Spectrom Ion Proc* 160: 339–356
51 Cohen S, Chait B (1997) Mass spectrometry of whole proteins eluted from sodium dodecyl
   sulfate-polyacrylamide gel electrophoresis gels. *Anal Biochem* 247: 257–267
52 Vorm O, Roepstorff P (1996) Detection bias gating for improved detector response and
   calibration in matrix-assisted laser desorption/ionization time-of-flight mass spectrometry.
   J Mass Spectrom 31: 351–358
53 Jensen ON, Larsen MR, Roepstorff P (1998) Mass spectrometric identification and micro-
   characterization of proteins from electrophoretic gels: strategies and applications. *Proteins:
   Structure, Function and Genetics* 999: 74–89

54 Oda Y, Chait B (1998) Purification and identification of phosphopeptides. Proc. 46th ASMS Conference of Mass Spectrometry and Allied Topics, Orlando, Florida, 30 May–4 June
55 Burlingame AL (1996) Characterization of protein glycosylation by mass spectrometry. *Curr Opin Biotechnol* 7: 4–10
56 Mørtz E, Sareneva T, Haebel S, Julkunen I, Roepstorff P (1996) Mass spectrometric characterization of glycosylated interferon-gamma variants separated by gel electrophoresis. *Electrophoresis* 17: 925–931
57 Mørtz E, Sareneva T, Julkunen I, Roepstorff P (1996) Does matrix-assisted laser desorption/ionization mass spectrometry allow analysis of carbohydrate heterogeneity in glycoproteins? A study of natural human interferon-gamma. *J Mass Spectrom* 31: 1109–1118
58 Olsen EHN, Rahbek-Nielsen H, Thøgersen IB, Roepstorff P, Enghild JJ (1998) Post-translational modifications of human inter-α-inhibitor: identification of glycans and disulfide bridges in heavy chains 1 and 2. *Biochemistry* 37: 408–416
59 Ploug M, Rahbek-Nielsen H, Nielsen PF, Reopstorff P, Danø K (1998) Glycosylation profile of a recombinant urokinase-type plasminogen activator receptor expressed in Chinese hamster ovary cells. *J Biol Chem* 273: 13933–13943
60 Krogh TN, Mirgorodskaya E, Mørtz E, Højrup P, Roepstorff P (1996) Characterization of protein glycosylation by post source decay and exoglycosidase digestion. Proc. 44th ASMS Conference of Mass Spectrometry and Allied Topics, Portland, Oregon, 12–16 May, 1336
61 Ens W, Standing K, Chernushevitc IV (eds) (1998) New methods for the study of biomolecular complexes. Kluver Academic Publishers, Dordrecht
62 Roepstorff P (1997) Mass spectrometry in protein studies from genome to function. *Curr Opin Biotech* 8: 6–13
63 Strupat K, Hillenkamp F (1998) Analysis of quaternary protein ensembles by matrix assisted laser desorption/ionization mass spectrometry. *J Am Chem Soc* 119: 1046–1052
64 Suckau D, Mak M, Przybylski M (1992) Protein surface-topology probing by selective chemical modification and mass spectrometric mapping. *Proc Natl Acad Sci USA* 89: 5630–5634
65 Ellison D, Hinton J, Hubbard SJ, Beinon RJ (1995) Limited proteolysis of native proteins: the interaction between avidin and proteinase K. *Prot Sci* 4: 1337–1345
66 Ploug M, Rahbek-Nielsen H, Ellis V, Roepstorff P, Danø K (1995) Chemical modification of the urokinase-type plasminogen activator and its receptor using tetranitromethane. Evidence for the involvement of specific tyrosine residues in both molecules during receptor-ligand interaction. *Biochemistry* 34: 12524–12534
67 Nordhoff E, Roepstorff P (1998) New methods for protein foot-printing using MALDI-MS. Proccedings of the 46th ASMS Conference on Mass Spectrometry and Allied Topics, Orlando, Florida, 30 May–4 June
68 Kussmann M, Bennett KL, Sørensen P, Godswon M, Bjoerk P, Roepstorff P (1998) Studying protein-protein interactions by protein cross-linking and differential MALDI mass spectrometric peptide mapping. Proceedings of the 44th ASMS Conference on Mass Spectrometry and Allied Topics, Orlando, Florida 30 May–4 June
69 Zhao Y, Muir TW, Kent SBH, Tischer T, Scardina JM, Chait BT (1996) Mapping protein-protein interactions by affinity-directed mass spectrometry. *Proc Natl Acad Sci USA* 93: 4020–4024
70 Schriemer DC, Li L (1996) Combining avidin-biotin chemistry with matrix-assisted laser desorption/ionization mass spectrometry. *Anal Chem* 68: 3382–3387
71 Girault S, Sagan S, Bolbach G, Lavielle S, Chassaing G (1996) The use of photolabeled peptides to localize the substance-P-binding site in the human neurokin-1 tachykinin receptor. *Eur J Biochem* 240: 215–222
72 Nelson RW, Krone JR, Bieber AL, Williams P (1995) Mass spectrometric immunoassay. *Anal Chem* 67: 1153–1158
73 Krone JR, Nelson RW, Dogruel D, Williams P (1997) BIA/MS: interfacing biomolecular interaction analysis with mass spectrometry. *Anal Chem* 69: 4363–4368
74 Sönksen CP, Nordhoff E, Jansson Ö, Malmqvist M, Roepstorff P (1998) Combining MALDI mass spectrometry and biomolecular interaction analysis using surface plasmon resonance. *Anal Chem* 70: 2731–2736

Proteomics in Functional Genomics
ed. by P. Jollès and H. Jörnvall
© 2000 Birkhäuser Verlag Basel/Switzerland

The chemistry of protein sequence analysis

John E. Shively

Division of Immunology, Beckman Research Institute of the City of Hope, Duarte, CA 91010, USA

Summary. N-terminal sequence analysis by Edman chemistry continues to play an important role in the structural analysis of proteins and peptides. Improvements in the sensitivity of the method have been achieved mainly at the level of increasing the sensitivity of the on-line analysis of PTH amino acids by RP-HPLC (reverse phase high performance chromatography). Using microbore columns (0.8–1.0 mm), it is possible to run standards at the 0.5–1.0 pmol level and to sequence samples in the 1–5 pmol range. Due to constraints in current chromatographic methods, it is unlikely that further improvements in sensitivity will be achieved by this approach alone. Although alternative Edman reagents, including fluorescent chemistries, have promised to increase the sensitivity of sequencing into the low femtomole range, none of the methods have progressed into routine usage. These reagents and chemistries are critically evaluated in this review, and the problems which have prevented their further development discussed. Instrumental constraints are also considered. It is concluded that the development of more sensitive methods requires further research into both the chemistry and the instrumentation, and that alternative separation and detection methods may also play a role.

Introduction

The primary structural analysis of proteins usually proceeds in two steps, starting with the sequence analysis of the N-terminus and proceeding to the sequence analysis of internal peptides generated by proteolytic or enzymatic cleavage of the protein into fragments. Although Edman sequencing excels in both approaches, in terms of the total number of proteins analyzed, it has been largely displaced by mass spectrometric methods in the last 5 years. There are three basic reasons for this paradigm shift. The first is the need for greater sensitivity than Edman chemistry can provide. Edman chemistry currently requires 1–10 pmol of sample for adequate analysis, and although the starting amounts (i.e. for sample digestion) for mass spectrometry are about the same, the amounts analyzed are easily 10–100-fold less. The second is the need for speed. A single cycle (identification of one amino acid) of Edman chemistry requires 30–60 min, whereas peptide fragmentation takes only seconds in a mass spectrometer. Speed is important in analysis of complex mixtures of proteins [e.g. samples from one-dimensional (1D) or 2D gels]. The third is that mass spectrometric methods excel in identifying amino acid derivatives (e.g. posttranslational modifications), whereas Edman chemistry performs poorly in this aspect. Nonetheless, Edman chemistry still plays an important role in protein structural analysis for the following reasons. It is the only method which can help confirm the structure of an intact protein (specifically, the protein must have an unblocked N-terminus for this to be true). The N-terminal sequence of an intact protein is a valuable proof of a protein's structure. While such a sequence is unlikely to confirm the entire sequence, it can confirm a substantial portion of the sequence, and the data, unlike mass spectrometric analysis, are quantitative in nature. Although attempts have been made, mass spectrometry does poorly in the sequence analysis of intact proteins. Mass spectrometry can also deliver an accurate mass for the intact protein (assuming not too much microheterogeneity), a datum which is supportive but not confirmatory of a protein's structure. On the other hand, some types of microheterogeneity such as glycosylation, which interfere with or complicate mass spectrometric sequence analysis, do not affect the overall performance of Edman chemistry (i.e. except for the identification of the derivatized amino acids, all others are easily identified). Perhaps the most noteworthy aspect of Edman chemistry in today's protein sequence arena is the complementary nature of the analysis to mass spectrometry. Since the two methods rely on completely different methods for structure determination, the combination of the two methods allows for a more confident identification of a protein's sequence. With these facts in mind, this chapter reviews the current situation and possible problems in the use of Edman chemistry on a variety of sample sources.

1. Edman Chemistry

While the basics of Edman chemistry have changed little from the day of Edman [1], the sensitivity of the method has improved over six orders of magnitude (from the micromole to the picomole level). These improvements followed the development of the gas phase sequencer [2] and more sensitive chromatographic methods. The three steps of Edman chemistry, which include coupling, cleavage, and conversion are shown in Figure 1. In the first step, the protein or peptide sample is immobilized on a carrier, and coupled with the Edman reagent, phenylisothiocyanate (PITC), to form a phenythiocarbamyl (PTC) derivative. This step requires that the protein posseses an unblocked N-terminus. Many modifications are known to occur in nature at the N-terminus; chief among them are the N-acetyl derivatives found in many proteins. In addition, the N-terminus may become blocked during sample handling, thus leading to many protocols designed to prevent such a mishap. In the second step, the PTC amino acid derivative is cleaved from the protein by anhydrous trifluoroacetic acid (TFA), resulting in the formation of an anilinothiazolinone derivative (ATZ) which is delivered to the conversion flask and converted to the phenythiohydantoin derivative (PTH) which is subsequently analyzed by reversed-phase HPLC. The sample immobilization step is critical, since the

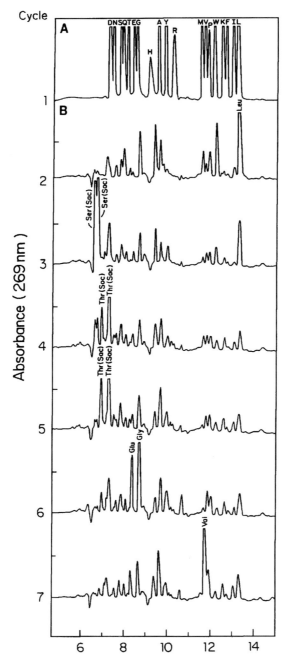

Figure 7. Sequence analysis of a glycoprotein on the Beckman LF 3600 sequencer. (A) The upper trace shows the separation of a 20-pmol PTH standard mixture. (B) The lower traces show the first seven cycles of the sequence analysis of 200 pmol of asialoglycophorin. Cycle 2 gives two peaks corresponding to O-glycosylated PTH-Ser, and cycles 3 and 4 give two peaks corresponding to O-glycosylated PTH-Thr.

and Thr. We have also identified PTH derivatives of glycosylated Asn residues in carcinoembryonic antigen [44]. However, as mentioned above, the most direct method for identification of these and other posttranslational modifications is mass spectrometry.

Since glycosylation, especially N-glycosylation, can complicate peptide mapping and subsequent sequence analysis of glycopeptides, it may be advisable to deglycosylate the sample before beginning the analysis. Two general methods are available. The first is the removal of N-glycosyl units with enzymes such as glycopeptidase F or N-glycopeptidase, the former for high-mannose-type structures, and the latter for complex-type structures. Both enzymes work best on denatured samples, and are compatible with SDS gel electrophoresis. After treatment, the SDS gel-separated band (from excess enzyme) can be subjected to in-gel trypsin digestion and peptide mapping (see chapter 3, this volume). It should be noted that N-glycopeptidase converts the previously glycosylated Asn residue to Asp. A chemical method using anhydrous TFMSA (trifluromethane sulfonic acid) and anisole (or thioanisole) for deglycosylation is also available [46]. This method cleaves all O-glycosidic bonds. In the case of N-linked carbohydrates, the pendant sugars will be trimmed back to the innermost GlcNAc attached to the asparagine. We have scaled this method down to the μg sample level [47, 48] and shown that the PTH-Asn-GlcNAc derivative can be identified on a gas phase sequencer as a distinct peak on LC analysis [48]. Further work is in progress to demonstrate this method as an in-gel method.

4. Future prospects

Modern Edman sequencers couple the chemistry of sequencing with the analysis of the PTH derivatives. While these instruments remain a critical tool for the protein chemist, there is still a need for greater sensitivity and speed. Improvements in chemistry and analytical methods have the potential to deliver both. However, current instruments handle relatively large (μl) volumes that are incompatible with the more sensitive analytical methods and which carry along unacceptably high levels of reagent and background peaks. The greatest hurdle to overcome now is the miniaturization of the instrument to match the nl volumes required in analytical methods such as CE or electroendosmotic flow LC. While no reports of such instruments are on the horizon yet, the technology for their construction exists. It is easy to predict that the next Edman sequencer will take advantage of chip technology.

Acknowledgments
The author acknowledges the use of data from Drs. Terry Lee, Kristine Swiderek, and William Henzel.

References

1 Edman P, Begg GA (1967) A protein sequenator. *Eur J Biochem* 1: 80–91
2 Hewick RM, Hunkapillaer MW, Hood LE, Dryer WJ (1981) A gas-liquid solid phase peptide and protein sequenator. *J Biol Chem* 256: 7990–7997
3 Matsudaira P (1987) Sequence for picomole quantities of proteins electroblotted onto polyvinylidene membranes. *J Biol Chem* 262: 10035–10038
4 Calaycay J, Rusnak M, Shively JE (1991) Microsequence analysis of peptides and proteins. An improved, compact, automated instrument. *Anal Biochem* 192: 23–31
5 Baumann M (1990) Comparative gas phase and pulsed liquid phase sequencing on a modified applied biosystems 477A sequencer. *Anal Biochem* 190: 198–208
6 Reim DF, Speicher DW (1993) High sensitivity gas phase sequence analysis of proteins and peptides on PVDF membranes using short cycle times. *Anal Biochem* 214: 87–95
7 Burkhart WA, Moyer MB, Bodnar WM, Everson AM, Valladares VG, Bailey JM (1995) Direct collection onto Zitex and PVDF for Edman sequencing: elimination of polybrene. In: Marshak D (ed) *Techniques in protein chemistry VI*, Academic Press, San Diego, 169–176
8 Tarr GE, Beecher JF, Bell M, McKean DJ (1978) Polyquaternary amines prevent peptide loss from sequenators. *Anal Biochem* 84: 622–627
9 Laursen RA (1971) Solid phase Edman degradation. *Eur J Biochem* 20: 89–102
10 Pappin DJC, Coull JM, Köster H (1990) Solid-phase sequence analysis of proteins electroblotted or spotted onto polyvinylidene difluoride membranes. *Anal Biochem* 187: 10–19
11 Pappin DJC, Coull JM, Koester H (1990) New approaches to covalent sequence analysis. *Current Res Prot Chem* 18: 191–202
12 Liang S-P, Laursen RA (1990) Covalent immobilization of proteins and peptides for solid-phase sequencing using prepacked capillary columns. *Anal Biochem* 188: 366–373
13 Ireland ID, Lewis DF, Li XF, Renborg A, Kwong S, Chen M, Dovichi NJ (1997) Double coupling Edman chemistry for high-sensitivity automated protein sequencing. *J Protein Chem* 16: 491–493
14 Sun T, Lovins RE (1972) Quantitative protein sequencing using mass spectrometry: use of low ionizing voltages in mass spectral analysis of methyl- and phenylthiohydantoin amino acid derivatives. *Anal Biochem* 45: 176–191
15 Rangarajan M, Ardrey RE, Darbre A (1973) Gas-liquid chromatography and mass spectrometry of amino acid thiohydantoins and their use in protein sequencing. *J Chromatogr* 87: 499–512
16 Bures EJ, Nika H, Chow DT, Morrison HD, Hess D, Aebersold R (1995) Synthesis of the protein-sequencing reagent 4-(3-pyridinylmethylaminocarboxypropyl) phenylthiohydantoins. *Anal Biochem* 224: 364–372
17 Hess D, Nika H, Chow DT, Bures EJ, Morrison HD, Aebersold R (1995) Liquid chromatography-electrospray ionization mass spectrometry of 4-(3-pyridinylmethylaminocarboxypropyl)phenylthiohydantoins. *Anal Biochem* 224: 373–381
18 Aebersold R, Bures EJ, Namchunk M, Goghari MH, Shushan B, Covey TC (1992) Design, synthesis, and characterization of a protein sequencing reagent yielding amino acid derivatives with enhanced detectability by mass spectrometry. *Protein Sci* 1: 494–503
19 Henry C (1998) Automated protein sequencers – alive and kicking. *Anal Chem News and Features* 401A–404A
20 Zhou J, Hefta S, Lee TD (1997) High sensitivity analysis of phenylthiohydantoin amino acid derivatives by electrospray mass spectrometry. *J Am Soc Mass Spectrom* 8: 1165–1174
21 Stolowitz ML, Paape BA, Dixit VM (1989) Thioacetylation method of protein sequencing: derivative of 2-methyl-5 (4H)-thiazolones for high-performance liquid chromatographic detection. *Anal Biochem* 181: 113–119
22 Stolowitz ML, Kim C-S, Marsh SR, Hood L (1993) Thiobenzoylation method of protein sequencing: gas chromatography/mass spectrometric detection of 5-acetoxy-2-phenyl-thiazoles. In: K Imahore, F Sakiyama (eds): *Methods in protein sequence analysis*, Plenum Press, New York, 37–44
23 Stolowitz M, Hood L (1993) Single syringe-pump solid-phase protein sequencer. In: R Hogue-Angeletti (ed) *Techniques in protein chemistry IV*. Academic Press, San Diego, 435–440

24 Tsugita A, Kamo M, Jone CS, Shikama N (1989) Sensitization of Edman amino acid derivatives using the fluorescent reagent, 4-aminofluorescein. *J Biochem* 106: 60–65

25 Farnsworth V, Steinberg K (1993) The generation of phenylthiocarbamyl or anilinothiazolinone amino acids from the postcleavage products of the Edman degradation. *Anal Biochem* 215: 200–210

26 Farnsworth V, Steinberg K (1993) Automated subpicomole protein sequencing using an alternative postcleavage conversion chemistry. *Anal Biochem* 215: 190–199

27 Totty NF, Waterfield MD, Hsuan JJ (1992) Accelerated high-sensitivity microsequencing of proteins and peptides using a miniature reaction cartridge. *Protein Sci* 1: 1215–1224

28 Slattery TK, Harkins RN (1993) *Analysis of complex protein mixtures on the HP-G1000A sequencer*. Academic Press, San Diego, 435–440

29 Li X-F, Hongji R, Lewis DF, Ireland ID, Waldron KC, Dovichi NJ (1997) Protein sequencing using microreactors and capillary electrophoresis with thermo-optical absorbance detection. In: DR Marshak (ed) *Techniques in protein chemistry VIII*. Academic Press, San Diego, 3–14

30 Li X-F, Waldron KC, Black J, Lewis D, Ireland I, Dovichi NJ (1997) Miniaturized protein microsequencer with PTH amino acid identification by capillary electrophoresis II. A syringe-pump-based system for covalent sequencing. *Talanta* 44: 401–411

31 Waldron KC, Li X-F, Chen M, Ireland I, Lewis D, Carpenter M, Dovichi NJ (1997) Miniaturized protein microsequencer with PTH amino acid identification by capillary electrophoresis I. An argon pressurized delivery system for adsorptive and covalent sequencing. *Talanta* 44: 383–399

32 Waldron KC, Dovichi NJ (1992) Sub-femtomole determination of phynylthiohydantoin-amino acids: capillary electrophoresis and thermooptical detection. *Anal Chem* 64: 1396–1399

33 Matsunaga H, Santa T, Lida T, Fukushima T, Homma H, Imai K (1996) Proton: a major factor for the racemization and the dehydration at the cyclization/cleavage stage in the Edman sequencing method. *Anal Chem* 68: 2850–2856

34 Tsunasawa S, Hirano H (1993) *Deblocking and subsequent microsequence analysis of N-terminally blocked proteins immobilized on PVDF membrane*. In: K Imahori, F Sakiyama (eds) *Methods in Protein Sequence Analysis*. Plenum Press, New York, 45–53

35 Gherorghe MT, Bergman T (1995) Deacetylation and internal cleavage of polypeptides for N-terminal sequence analysis. In: MZA, E Appella (ed): *Methods in protein structure analysis*. Plenum Press, New York, 81–86

36 Miyatake N, Kamo M, Satake K, Uchiyama Y, Tsugita A (1993) Removal of N-terminal formyl groups and deblocking of pyrrolidone carboxylic acid of proteins with anhydrous hydrazine vapor. *Eur J Biochem* 212: 785–789

37 Henschen A (1993) Identification of tyrosine sulfate and tyrosine phosphate residues during sequence analysis. *Protein Sci* 2: 152

38 Meyer HE, Hoffman-Posorske E, Korte H, Heilmeyer LMG Jr (1986) Sequence analysis of phosphoserine containing peptides. *FEBS Lett* 204: 61–66

39 Meyer HE, Hoffmann-Posorske E, Heilmeyer LMG Jr (1991) Determination and location of phosphoserine in proteins and peptides by conversion to S-ethylcysteine. *Methods Enzymol* 201: 169–185

40 Annan WD, Manson W, Nimmo JA (1982) The identification of phosphoseryl residues during the determination of amino acid sequence in phosphoproteins. *Anal Biochem* 121: 62–68

41 Parten BF, McDowell JH, Nawrocki JP, Hargrave PA (1994) Characterization of phosphorylation sites in bovine rhodopsin using modified gas phase sequencing programs. In: *Techniques in protein chemistry V*. Academic Press, San Diego, 159–166

42 Wang Y, Fiol CJ, DePaoli-Roach AA, Bell AW, Hermodson MA, Roach PJ (1988) Identification of phosphorylation sites in a peptides using a gas phase sequencer. *Anal Biochem* 174: 537–547

43 Meyer HE, Hoffmann-Posorske E, Heilmeyer LMG Jr (1991) Sequence analysis of phosphotyrosine containing peptides. *Methods Enzymol* 201: 206–224

44 Lin X, Wulf L, Khan S, Ford CF, Swiderek KM (1997) Positive identification of glycosylation sites in proteins and peptides using a modified Beckman LF 3600 N-terminal protein sequencer. In: DR Marshak (ed) *Techniques in protein chemistry*. Vol. 8, Academic Press, San Diego, 331–339

45 Pisano A, Packer NH, Redmond JW, Williams KL, Gooley AA (1995) Identification and characterization of glycosylated phenythiohydantoin amino acids. In: MZA, E Appella (ed) *Methods in protein structure analysis.* Plenum Press, New York, 69–80

46 Edge AB, Faltynek CR, Hof L, Reichert LE Jr, Weber P (1981) Deglycosylation of glycoproteins by trifluoromethanesulfonic acid. *Anal Biochem* 118: 131–137

47 Hefta SA, Paxton RJ, Shively JE (1990) Sequence and glycosylation site identity of two distinct glycoiforms of nonspecific crossreacting antigen as demonstrated by sequence analysis and fast atom bombardment mass spectrometry. *J Biol Chem* 265: 8618–8626

48 Paxton RJ, Mooser G, Pande H, Lee TD, Shively JE (1987) Sequence analysis of carcinoembryonic antigen: identification of glycosylation sites and homology with the immunoglobulin supergene family. *Proc Natl Acad Sci USA* 84: 920–924

Proteomics in Functional Genomics
ed. by P. Jollès and H. Jörnvall
© 2000 Birkhäuser Verlag Basel/Switzerland

The alkylated thiohydantoin method for C-terminal sequence analysis

David R. Dupont[1], MeriLisa Bozzini[1] and Victoria L. Boyd[2]

[1] *PE Biosystems, 700 Lincoln Centre Drive, Foster City, California 94404, USA and*
[2] *CELERA, 850 Lincoln Centre Drive, Foster City, California 94404, USA*

Summary. The alkylated-thiohydantoin method for C-terminal sequencing makes a significant improvement to the thiohydantoin method first described by Schlack and Kumpf. Prior to cleavage from the protein, the C-terminal thiohydantoin is alkylated, making it a better leaving group than the unmodified thiohydantoin. The C-terminal alkylated-thiohydantoin can be cleaved from the protein under conditions that simultaneously form the next thiohydantoin. Combining cleavage and thiohydantoin formation in one step eliminates the need for activating the C-terminal carboxyl group before every sequencing cycle and prevents detection of C-termini formed by random cleavage of peptide bonds in the protein during the sequencing chemistry. The alkylated-thiohydantoin method includes the presequencing modification of cysteine and lysine and the automated modification of aspartic and glutamic acids, serine and threonine. Modifying the reactive side-chain groups improves the ability to sequence through and detect these amino acids. The alkylated-thiohydantoin method can sequence through and detect 19 of the 20 genetically coded amino acids. Sequencing stops at proline residues.

Introduction

The development of a chemical method for C-terminal sequence analysis of proteins which is as sensitive and robust as Edman chemistry remains an unattained goal. More than 20 years before Edman published his method [1] for the sequential degradation of a peptide chain from the N-terminus, Schlack and Kumpf published the first C-terminal degradation procedure [2]. Their method, consisting of converting the C-terminal amino acid of a peptide or protein to an acylthiohydantoin followed by cleavage of the thiohydantoin from the parent molecule, is typically referred to as the thiohydantoin (TH) method. Although a number of different C-terminal degradation chemistries have been proposed, the TH method remains the most studied and is currently the most widely used method for C-terminal sequence analysis. (For a review of C-terminal sequencing methods, see Inglis [3]).

The TH method has several drawbacks when compared with Edman chemistry. The TH derivative formed from the C-terminal amino acid of the protein is relatively stable; unlike the cyclic anilinothiazolinone (ATZ) derivative formed from the N-terminal amino acid during cleavage in Edman chemistry, TH formation does not promote cleavage of the C-terminal amino acid from the protein. The relatively harsh reagents and conditions necessary to cleave the TH can cause modification of reactive amino acid side chains and cleavage of the peptide backbone of the protein.

Fragmentation of the protein reduces the potential length of the sequencing run and makes data interpretation more difficult due to TH derivatives resulting from the newly formed C-termini. During each cycle of sequence analysis, the new C-terminal amino acid must be activated before a TH can be formed. This increases the complexity of the chemistry, introduces the opportunity for side reactions with the C-terminal amino acid or with reactive side chains of other amino acids and reduces the overall chemical efficiency.

1. The alkylated thiohydantoin method

The alkylated-thiohydantoin (ATH) method for C-terminal sequence analysis makes a significant improvement to the TH method. Published in 1992 [4], a unique feature of the ATH-sequencing method is S-alkylation of the TH formed at the C-terminus prior to cleavage from the protein. The resulting ATH is a better leaving group than the unmodified TH. The ATH can be cleaved from the protein with {NCS}$^-$ under conditions that simultaneously derivatize the new C-terminal amino acid to form a TH. Because the ATH can be displaced by {NCS}$^-$, cleavage and TH formation are combined in one step, eliminating the need for activation of the C-terminal carboxyl group before every sequencing cycle. Elimination of repeated carboxyl group activation simplifies sequence interpretation significantly, because any C-termini formed by random cleavage of peptide bonds in the protein during sequence analysis cannot form TH. This is a key advantage of the ATH method, preventing the detection of amino acid background resulting from fragmentation of the protein sample. Each subsequent alkylation of the proteinyl-TH, followed by cleavage/derivatization with {NCS}$^-$, comprises one cycle of sequence analysis. The steps of the ATH method are outlined in Figure 1.

Automated C-terminal protein sequence analysis using the ATH method has provided structural information that could not be obtained, or confirmed information that was difficult to obtain, by other methods of analysis [5–7]. Over 50 recombinant and natural proteins with a wide range of molecular weights (7.6–97 kDa) have been analyzed using either the ABI Model 477 or the Procise instrument platforms. Up to eight residues of C-terminal sequence were obtained, and usually at least four residues were detected for most proteins, including glycosylated proteins. Sequence truncations and heterogeneity were identified from a number of proteins [7].

The ATH sequencing method is run using the Procise sequencer and C-terminal sequencing reagents supplied by PE Biosystems. The ATH method is compatible with conventional methods of protein sample preparation, including the use of polyvinylidene fluoride (PVDF) membranes for electroblotting and as a sequencing support. The reagents and solvents used are listed in Table 1.

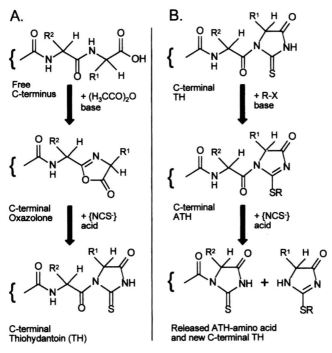

Figure 1. The alkylated-thiohydantoin method. (A) Activation and thiohydantoin formation. (B) Thiohydantoin alkylation and cleavage/thiohydantoin formation.

Table 1. C-terminal sequencing reagents

Bottle position	C-terminal reagent	Purpose
C1	alkylated thiohydantioin (ATH) standard	calibrate sequencing results
C2	unused	
C3	methylimidazole in acetonitrile	catalyze acetylation of OH of Ser/Thr
C4	piperidine thiocyanate in acetonitrile	form piperidine amide of Asp/Glu
C5	acetonitrile	wash flask
C6	acetic anhydride/lutidine in acetonitrile	activate C-terminus, acetylate Ser/Thr
C7	ethyl acetate	wash reaction cartridge and valve blocks, transfer ATH-AA to flask
C8	2-(bromomethyl)naphthalene in acetonitrile	alkylate newly formed thiohydantoin
C9	20% acetonitrile in water	dissolve flask contents, transfer to HPLC
C10	tetrabutylammonium thiocyanate in acetonitrile	form C-terminal thiohydantoin
C11	2% diisopropylethylamine in heptane	provide basic environment for alkylation
C12	trifluoroacetic acid	generate {NCS}$^-$ to cleave ATH-AA, form TH-AA

The ATH method can sequence through and detect 19 of the 20 gene-tically coded amino acids. Sequence analysis stops at proline residues.

1.1. Activation

In the ATH-sequencing method, activation of the free carboxylic acid group at the C-terminus occurs only once in order to form the first proteinyl-TH. A mixture of acetic anhydride (Ac_2O) and lutidine is used for carboxyl group activation, resulting in the formation of a C-terminal oxazolone [8]. The C-terminal oxazolone can react with excess activating reagent under strongly basic conditions. O-Acetylation of an oxazolone, the first step of the Dakin-West reaction [9], is promoted in the presence of a strong base or an acylation catalyst. Because the O-acetylated oxazolone reacts very slowly (relative to the unmodified oxazolone) to form a TH, the sequencing initial yield is reduced if the first step, or any subsequent step, of the Dakin-West reaction occurs. Therefore, lutidine, a weak base and poor acylation catalyst, is used during activation with Ac_2O.

1.2. Thiohydantoin formation

After initial activation of the C-terminal carboxyl group of a protein, the C-terminal amino acid is derivatized into a TH by thiocyanate anion ($\{NCS\}^-$). Tetrabutylammonium thiocyanate in acetonitrile is delivered to the sequencer reaction cartridge together with trifluoroacetic acid (TFA) vapor to form the C-terminal TH. The quarternary ammonium salt of $\{NCS\}^-$ was selected as the source of thiocyanate because of its enhanced stability in solution and because the quaternary ammonium counterion is nonnucleophilic, thereby minimizing potential side reactions. The pro-teinyl-TH is stable in the presence of the reagents used for TH formation: Ac_2O and lutidine, and later, $\{NCS\}^-$ and TFA. Figure 1A depicts the initial activation and TH formation.

After alkylation of the TH, treatment of the sample with tetrabutylam-monium thiocyanate in acetonitrile together with TFA vapor cleaves the ATH from the protein and forms a TH from the newly resulting C-terminal amino acid. Successive amino acids are removed from the C-terminus of the protein by repeating the alternating alkylation and cleavage/thiohydan-toin formation steps, as shown in Figure 1B.

1.3. Thiohydantoin alkylation

The TH is treated with 2-(bromomethyl)naphthalene under basic condi-tions to form the ATH. Benzyl bromide analogs in general and 2-(bromo-

2.5. Effects on other amino acids

During the course of derivatization of Cys, Lys, Asp, Glu, Ser and Thr residues, other reactive amino acids can be modified. PIC reacts with the hydroxyl group of Tyr to form an arylcarbamate, and thereby minimizes acetylation of Tyr during exposure to Ac_2O. Arylcarbamates are readily hydrolyzed, so Tyr residues are detected as both free Tyr-ATH and acetylated-Tyr-ATH. The sulfhydryl group of unmodified Cys reacts with PIC to form a thiocarbamate, which can eliminate during the sequencing chemistry to form dehydroalanine. PIC pretreatment alone does not improve the ability to detect Cys residues. Under the conditions for modification of Ser and Thr hydroxyl groups, Ac_2O also reacts with Arg, His and Lys if the ε-amino group has not been derivatized with PIC.

3. Future developments

There are certain isothiocyanate reagents that can both activate the C-terminus and form thiohydantoin. Because both steps can be done with a single reagent, this should minimize the possibility of side reactions and improve sequence analysis of initial yields. Several publications [13, 14] describe the application of one such reagent, diphenylphosphoroisothiocyanatidate (DPP-ITC), to C-terminal sequence analysis. Recent publications [15, 16] detail the use of acetyl isothiocyanate (Ac-ITC) for C-terminal activation and thiohydantoin formation. It is proposed that these reagents can form a thiohydantoin without proceeding through an oxazolone, which should lessen the potential for side reactions and improve the sequence analysis of the initial yield.

Figure 6 shows cycle 1 from 1 nmol of horse apomyoglobin applied to a PVDF membrane and analyzed for sequence using different activation reagents. In Figure 6A, using the current activation with Ac_2O, the initial yield was 170 pmol or 17%, within the typical initial yield range of 15–25% for horse apomyoglobin. In Figure 6B, using DPP-ITC for initial activation, the initial yield was 460 pmol or 46%. In Figure 6C, using Ac-ITC for initial activation, the initial yield was 270 pmol or 27%.

Both DPP-ITC and Ac-ITC show potential for improvements of the sequencing initial yields compared with our current results using acetic anhydride for activation. We will continue to optimize the on-instrument reaction conditions for these reagents. To better compare the relative benefits and weaknesses, we will be testing the different activation strategies on a set of model peptides with each of the common amino acids represented at the C-terminus.

Acknowledgments
We thank Drs Cynthia Wadsworth, Richard L. Noble, Thomas Coleman, Ken Graham and William E. Werner, and B. John Bergot for their support in the development of this method.

A. Acetic anhydride activation

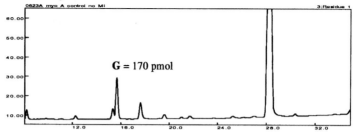

G = 170 pmol

B. DPP-ITC activation

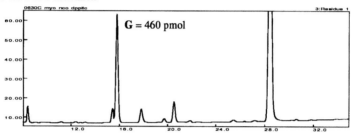

G = 460 pmol

C. Ac-ITC activation

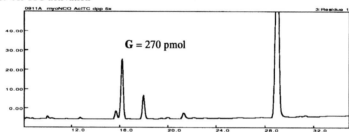

G = 270 pmol

Figure 6. Comparison of sequencing results from 1 nmol of horse apomyoglobin applied to PVDF membrane and activated using (A) acetic anhydride, (B) DPP-ITC, (C) Ac-ITC.

References

1 Edman P (1950) Method for the determination of the amino acid sequence of peptides. *Acta Chem Scand* 4: 283–293
2 Schlack P, Kumpf W (1926) Über eine neue Methode zur Ermittlung der Konstitution von Peptiden. *Z Physiol Chem* 154: 125–170
3 Inglis AS (1991) Chemical procedures for C-terminal sequencing of peptides and proteins. *Anal Biochem* 195: 183–196
4 Boyd VL, Bozzini M, Zon G, Noble RL, Mattaliano RJ (1992) Sequencing of peptides and proteins from the carboxy terminus. *Anal Biochem* 206: 344–352
5 Martinez A, Knappskog PM, Olafsdottir S, Doskeland AP, Eiken HG, Svebak RM, Bozzini M, Apold J, Flatmark T (1995) Expression of recombinant human phenylalanine hydroxylase as fusion protein in *Escherichia coli* circumvents proteolytic degradation by host cell proteases. Isolation and characterization of the wild-type enzyme. *Biochem J* 306: 589–597

6 Lu KV, Rohde MF, Thomason AR, Kenney WC, Lu HS (1995) Mistranslation of a TGA termination codon as tryptophan in recombinant platelet-derived growth factor expressed in *Escherichia coli. Biochem J* 309: 411–417

7 Bozzini M, Zhao J, Yuan P-M, Ciolek D, Pan Y-C, Horton J, Marshak DR, Boyd VL (1995) Applications using an alkylation method for carboxy-terminal sequencing. In: J Crabb (ed) *Techniques in protein chemistry VI.* Academic Press, San Diego, 229–237

8 Boyd VL, Bozzini M, Guga PJ, DeFranco RJ, Yuan P-M (1992) Activation of the carboxy terminus of a peptide for carboxy-terminal sequencing. *J Org Chem* 60: 2581–2587

9 Buchanan GL (1988) The Dakin-West reaction. *Chem Soc (London) Reviews* 17: 91–109

10 Stark GR (1968) Sequential degradation of peptides from their carboxyl termini with ammonium thiocyanate and acetic anhydride. *Biochemistry* 7: 1796–1807

11 Brune DC (1992) Alkylation of cysteine with acrylamide for protein sequence analysis. *Anal Biochem* 207: 285–290

12 Blagbrough IS, Mackenzie NE, Ortiz C, Scott AI (1986) The condensation reaction between isocyanates and carboxylic acids. A practical synthesis of substituted amides and anilides. *Tetrahedron Lett* 27: 1251–1254

13 Kenner GW, Khorana HG, Stedman RJ (1953) Peptides. Part IV. Selective removal of the C-terminal residue as a thiohydantoin. The use of diphenylphosphoroisothiocyanatidate. *Chem Soc J (London)*, 673–678

14 Bailey JM, Nikfarjam F, Shenoy NR, Shively JE (1992) Automated carboxy-terminal sequence analysis of peptides and proteins using diphenylphosphoroisothiocyanatidate. *Protein Sci* 1: 1622–1633

15 Anumula KR, Tang S (1995) Novel chemistry for sequencing of proteins from the carboxyl terminus yields a simple method. *FASEB J* 9: A1477

16 Mo B, Li J, Liang S (1997) Chemical carboxy-terminal sequence analysis of peptides using acetyl isothiocyanate. *Anal Biochem* 252: 169–176

Proteomics in Functional Genomics
ed. by P. Jollès and H. Jörnvall
© 2000 Birkhäuser Verlag Basel/Switzerland

Ladder sequencing

Tomas Bergman

*Department of Medical Biochemistry and Biophysics, Karolinska Institutet,
SE-171 77 Stockholm, Sweden*

Summary. Ladder sequencing of polypeptides involves progressive N- or C-terminal amino acid truncation via chemical or enzymatic treatments. Peptide ladders are generated in which each component differs from the next by one residue. The ladder components are analyzed by mass spectrometry, and the amino acid sequence is deduced from the mass differences between consecutive fragments. Chemical procedures are common in N-terminal degradation, whereas proteolytic digestion is often used in C-terminal sequence analysis. Matrix-assisted laser desorption/ionization mass spectrometry is widespread for one-step readout of the peptide ladders and provides high sensitivity in combination with robustness and ease of use. The particular advantage of ladder sequencing in relation to other techniques for sequence analysis is the high data acquisition rate and the very good sample throughput that can be achieved. Multiple determinations are carried out within minutes at high sensitivity and low sample consumption. Several reports demonstrate analysis at the low picomole to femtomole level.

Introduction

Sequence determination of polypeptides via the generation of peptide ladders and analysis by mass spectrometry (MS) has emerged as a rapid analytical approach. The principle is progressive truncation of the original peptide from either the N- or the C-terminal end to produce a series of fragments, a peptide ladder, in which each member differs from the next by one amino acid residue. The molecular masses of the fragments generated are subsequently determined in a single mass spectrometry step, and the mass difference between each component in the ladder from high to low mass assigns both the identity of the amino acid residues and their sequence in the original peptide (Fig. 1). The ladder peptides can be formed by chemical degradation or via proteolytic digestion with exopeptidases. A majority of the protocols presented for N-terminal degradation are focused on chemical procedures, whereas enzymatic techniques with carboxypeptidases (CPs) are common for C-terminal sequence analysis (Tab. 1). For MS and readout of the peptide masses in the ladders generated, both fast atom bombardment (FAB), plasma desorption (PD) and electrospray (ES) ionization techniques have been employed, but the most widespread MS approach is matrix-assisted laser desorption/ionization (MALDI) with time-of-flight (TOF) detection due to high sensitivity in combination with relatively good salt tolerance (Tab. 1).

POLYPEPTIDE
SAMPLE

CHEMICAL OR ENZYMATIC
LADDER GENERATING
PROCESS

N-term   C-term

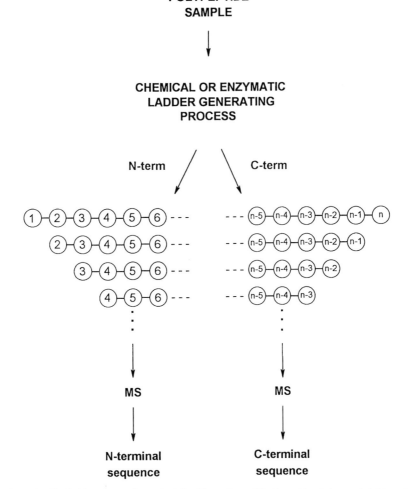

MS   MS

N-terminal   C-terminal
sequence   sequence

Figure 1. The ladder-sequencing principle. The polypeptide (*n* residues) is nonstoichiometri-
cally degraded from the N- or the C-terminus, generating a population of successively truncated peptides that are analyzed in a single MS step. The mass differences between consecutive
components in the peptide ladder assign both the identity of residues and the amino acid
sequence.

1. Chemical degradation

Chemical ladder-generating procedures have mainly been developed for
N-terminal approaches, but also a few C-terminal protocols have been
presented (cf. Tab. 1). N-terminal ladder chemistries are often based on the
classical Edman degradation method [1]. In this procedure the protein or
peptide is first reacted at its N-terminal α-amino group with phenylisothiocyanate (PITC), resulting in the formation of a peptidyl phenyl-

Table 1. Ladder sequencing protocols[a] (for abbreviations, cf. text)

Direction	Type	Degradation	MS	Amount[b]	Residues[c]	Reference
N	chem	phenylisothio-cyanate + phenyliso-cyanate	MALDI	200 pmol – 10 nmol	11	[2]
N	chem	trifluoroethyl-isothiocyanate	MALDI	1 – 20 pmol	7	[3]
N	chem	allyl isothio-cyanate	MALDI	10 – 30 pmol	7	[5]
C	chem	pentafluoro-propionic acid or heptafluoro-butyric acid	FAB or ES	100 – 200 pmol	8	[6]
C	chem	perfluoroacyl anhydride	FAB	1 – 10 nmol	18	[7]
C	enz	CPY and CPM II	PD	100 pmol – 1 nmol	10	[9]
C	enz	CPY	PD	100 pmol	4	[12]
C	enz	CPP	ES	2 nmol	19	[14]
C	enz	CPP	ES	1 – 2 nmol	6	[15]
C	enz	CPY and CPB	MALDI	500 pmol	8	[10]
C	enz	CPY	MALDI	5 – 500 pmol	19	[16]
C	enz	CPY and CPP	MALDI	500 pmol	11	[11]
C	enz	CPY and CPP	MALDI	500 pmol	20	[17]
C	enz	CPY and CPB	MALDI	20 – 40 pmol	5	[18]
C/N[d]	enz	CPY/LAP	FAB	500 pmol – 5 nmol	8/4[d]	[19]
C/N	enz	CPP/APM	MALDI	0.5 pmol	5/3	[20]
C	enz/[e] chem	1. CPP 2. a) acetic anhydride b) guanidine thiocyanate c) ammonium hydroxide	MALDI	10 – 30 pmol	8	[21]

[a] The purpose is to give an idea about available techniques rather than to provide a complete coverage of the field.
[b] Typical polypeptide amounts required for analysis.
[c] Reported number of consecutive residues sequenced.
[d] Both C- and N-terminal degradation employed, in which case the first number of residues applies to the C-terminal degradation and the second to the N-terminal degradation.
[e] Application of an enzymatic protocol was followed by a chemical procedure.

thiourea. Cleavage of the N-terminal residue is effected by anhydrous acid, normally trifluoroacetic acid (TFA), which causes cyclization and concomitant release of the five-membered anilinothiazolinone (ATZ) derivative. The ATZ amino acid is extracted and converted to the more stable phenylthiohydantoin (PTH) derivative by treatment with aqueous acid (TFA). The PTH amino acid is identified via a reverse-phase HPLC (high performance liquid chromatography) procedure.

Chait et al. [2] early described a modified Edman approach involving a terminating agent, phenylisocyanate (PIC), to produce the ladder peptides for MS readout and analysis of the N-terminal sequence. Instead of using only PITC for the coupling step in a manual procedure, a mixture of PITC and PIC at a ratio of typically 20 : 1 (v/v) was used. The result is that a fraction of the available N-termini in each cycle of the ladder-sequencing process will couple to PIC and is consequently not amenable to cleavage with TFA and further processing. Degradation is thus terminated for this peptide population, and the net result is an accumulation of representatives of each N-terminally truncated species of the original polypeptide over a number of residues. The peptides that are not blocked by PIC will couple to PITC and proceed normally through the Edman cycle. The set of nested peptides generated, i.e. the peptide ladder, is analyzed by MALDI-MS, which allows subpicomole detection at a mass accuracy of up to 100 ppm and a practical upper limit for the starting peptide at approximately 50–60 residues [2]. In this and in most other reports where MALDI is used to readout the peptide ladder, α-cyano-4-hydroxycinnamic acid (CHCA) in acetonitrile/TFA is used as matrix, which enables the use of ultraviolet (UV) laser equipment, e.g. nitrogen laser light at 337 nm. Chait et al. [2] demonstrated N-terminal sequence determination for up to 11 residues for peptides in the range 10–20 residues, one of which was phosphorylated where the site of this modification could be detected via the ladder-sequencing technique. The consumption of peptide sample was in the range 200 pmol–10 nmol (Tab. 1), which is high in relation to the actual analytical sensitivity and mainly explained by losses of material during the washes necessary to remove excess PITC/PIC and other reagents [2]. The sample is contained in an Eppendorf tube and not specifically immobilized to a surface during degradation, which makes it difficult to retain enough sample for analysis starting from low picomole amounts.

Bartlet-Jones et al. [3] took the concept with a modified manual Edman degradation further by introducing a novel, volatile coupling reagent, trifluoroethylisothiocyanate (TFEITC). The idea is that all reagents, solvents and by-products of the chemistry should be volatile and removable by vacuum. In this manner extractions could be eliminated, and the risk for work-up peptide losses in theory minimized. However, the key reagent TFEITC is not commercially available and has to be synthesized using a laborious procedure [3]. The ladder-sequencing protocol involves generation of a nested set of peptides by addition of equal aliquots of starting peptide in each cycle and driving both the coupling and the cleavage reactions to completion. No chain-terminating agents are employed, and elaborate washing procedures are not required. Each cycle takes 35–40 min [3]. In addition to TFA, heptafluorobutyric acid (HFBA) was used for the cleavage reaction. The mass spectrometric analysis was by MALDI-TOF employing CHCA as matrix. In relation to the Chait et al. procedure [2], the benefits are mainly twofold: less peptide material is necessary since extractive losses are eliminated, and the presence of a free N-terminal amine in each component of the peptide ladder improves the ionization efficiency. The amine can be further employed for modification with a sensitivity-enhancing agent such as quaternary ammonium alkyl active esters [3]. Bartlet-Jones et al. [3] succeeded in performing a ladder sequence analysis through six cycles of a 14-residue peptide, consuming a total of only 700 fmol starting peptide. The longest sequence reported was seven residues, and the peptide amounts necessary varied in the range of 1–20 pmol (Tab. 1). The Bartlet-Jones et al. protocol has been applied to sequence analysis of novel β-lactoglobulins [4].

Further development of the concept with volatile reagents for the N-terminal ladder generating chemistry was made by Gu and Prestwich [5]. In their protocol allyl isothiocyanate (AITC) replaces TFEITC as the volatile amine-modifying reagent. In contrast to the latter substance, AITC is commercially available (Aldrich) and stable up to 80°C. However, it is only about half as reactive as PITC with primary amines (TFEITC is in contrast about twice as reactive as PITC using phenylalaninamide as test substance) [5]. The outline of the procedure is similar to that of Bartlet-Jones et al. [3] with addition of fresh peptide at each cycle to generate the N-terminal ladder for MALDI-MS readout. Cleavage of residues is accomplished with HFBA, and the MALDI-TOF analysis is performed using CHCA or sinapinic acid (SA) matrices. SA was employed for a 33-residue peptide, the longest tested, whereas CHCA was used for the rest of the peptides that ranged down to eight residues [5]. The largest number of successful cycles yielding N-terminal sequence information was seven, as was also the case in the Bartlet-Jones et al. report (Tab. 1), whereas the consumed peptide amounts were a little bit higher, in the range 10–30 pmol [5].

Regarding C-terminal ladder-generating chemistries, an interesting protocol has been developed by Tsugita et al. [6] employing hydrolysis of peptides (from 6 residues) and a small protein (lysozyme, 129 residues) by vapors of fluorinated carboxylic acids. Pentafluoropropionic acid and heptafluorobutyric acid at 90% (v/v) in water was found to be more useful than TFA to generate the C-terminal ladders (Tab. 1). The dried sample (100–200 pmol in a glass tube) was contacted by vapors formed at 90°C from the aqueous acid for typically 2–4 h. The dried hydrolyzate was dissolved in aqueous acetic acid and mixed with a matrix consisting of glycerol and thioglycerol for analysis by FAB-MS, or dissolved in acetic acid/aqueous methanol for ES-MS. Cysteine residues in lysozyme were pyridylethylated via alkylation with vinylpyridine to prevent disulfide interaction during and after acid degradation [6]. The longest sequence stretch reported was eight residues starting from 125 pmol magainin 1 (23 residues), employing 90% pentafluoropropionic acid in the vapor phase for 2 h, and ES-MS for analysis of the C-terminal peptide ladder. A complicating factor in the ES mode is the large number of multiply charged peptide components that makes it difficult to interpret mass spectra and to readout the ladder. A consequence is that mass spectrometric resolution becomes a critical issue: it must be high enough to be able to judge the charge state of the different ions in the spectra in order to determine the ladder-forming mass values. On the other hand FAB, as well as PD and MALDI, normally generates predominantly singly charged peptide ions (molecular ions, $[M + H]^+$) and thus the spectra become less complex and more easy to interpret. Tsugita et al. [6] suggest a mechanism for the acid degradation involving selective formation of an oxazolone ring at the C-terminal amino acid followed by hydrolytic removal of this residue. Side reactions were also observed, mainly peptide bond cleavage at the C-terminal side of internal aspartic acid residues and at the N-terminal side of serine residues.

Takamoto et al. [7] reported on a follow-up study to the Tsugita et al. procedure [6]. In this protocol, the anhydrides corresponding to the previously used acids are employed, i.e. pentafluoropropionic anhydride and heptafluorobutyric anhydride [7]. The peptide sample is dried in a glass tube and exposed to the perfluoroacyl anhydride vapor at $-20°C$ for 0.5–1 h. Following the reaction, the product is exposed to pyridine/water vapor at 100°C for 30 min. The readout of the peptide ladders was by FAB-MS, both in the positive and in the negative ion mode. The latter is helpful to maximize the amount of sequence information obtained, since positive charges can get lost in the peptides degraded [7]. However, the positive ion mode is generally more sensitive than the negative ion mode. Takamoto et al. [7] suggest that the sequential C-terminal degradation proceeds via active intermediates, e.g. oxazolones. Applying pentafluoropropionic anhydride to a 23-residue peptide revealed a FAB mass spectrum containing C-terminal sequence information for 18 residues (Tab. 1). The amount of peptide employed in these degradations varied in the range 1–10 nmol [7].

2. Enzymatic degradation

The standard method for exopeptidase sequencing is to collect aliquots of the reaction mixture at different time points, quenching the enzymatic reaction, and then to perform analysis. Traditionally this has been carried out by direct determination of the released amino acids, which is complicated by amino acid contamination in the enzyme solutions, by enzyme autolysis, and by lack of sensitivity under the conditions possible to use. However, when exopeptidase digestion is combined with MS, direct analysis of the resulting mixture of truncated peptides and readout of the peptide ladders is possible.

Frequently used carboxypeptidases are CPY, CPP, CPMII and CPB, and commonly employed aminopeptidases (APs) are leucine AP (LAP) and APM (Tab. 1). CPY, isolated from yeast [8], is a popular enzyme for sequence determination, since it penetrates all amino acid sequences, including those with proline. However, CPY still has a preference for hydrophobic residues (except proline), whereas hydrophilic residues, in particular glycine and aspartic acid in the penultimate position and basic amino acids in the ultimate position, are more slowly released. CPMII from malt also releases essentially all amino acids, but with a preference for arginine and lysine, and could thus complement CPY in sequence analysis [9]. Furthermore, CPB from pancreas gives hydrolysis at basic amino acids (lysine and arginine) in the C-terminal position and is therefore of value in combination with CPY [10]. CPP from *Penicillium* is fairly unspecific and can release many residues including proline, aspartic acid and glutamic acid, whereas the release of serine and glycine is considerably retarded [11]. Regarding aminopeptidases, both LAP and APM from kidney have long been used for N-terminal sequence analysis. A difference is that the amido group of asparagine or glutamine is hydrolyzed by LAP but not by APM. LAP cannot penetrate sequences with lysine and arginine, and proline in the penultimate position stops LAP hydrolysis. The latter is also valid for APM.

Analysis using ^{252}Cf PD-MS to identify the truncated peptides resulting from exopeptidase digestion of polypeptides in the range 900–3500 Da was studied by Klarskov et al. [9]. C-terminal sequence analysis for up to 10 residues was achieved using a combination of CPY and CPMII (Tab. 1). Samples in the low picomole range were digested *in situ* on a nitrocellulose target. In this case the molecular weight could first be determined with the same material. Digestion in solution required peptide amounts at the high picomole to low nanomole level. The *in situ* approach generated only a limited amount of sequence information, the C-terminal or a few C-terminal residues, whereas extended sequence information necessitated time-course studies using digestion in solution [9]. A general problem with exopeptidases for the ladder-generating process is that the different rates at which peptide bonds are hydrolyzed depends on the actual structure of the polypeptide. This makes it sometimes difficult to identify all components of the peptide ladder. Such a situation is the case when a slowly released residue is followed by a rapidly released residue, and the ladder component having the rapidly released amino acid at its C-terminus may be below the detection limit of the MS system. In other situations the release of a certain residue may be so slow that in practical terms it can be regarded as refractory to digestion. To overcome similar problems, Klarskov et al. [9] mixed CPY and CPMII or used the enzymes in succession with heat inactivation in between. Wang et al. [12] investigated the reaction conditions for CPY *in situ* digestion on the nitrocellulose foil in PD-MS and applied the results to determination of a cleavage site in the amyloid precursor protein. Continuous peptide ladders defining up to four residues and starting from 100 pmol β/A4 peptide (16 residues) were obtained (Tab. 1).

Analysis of exopeptidase digests via ES-MS is sensitive, and the technique has recently undergone several improvements, e.g. the introduction of use of nanoelectrospray [13]. An interesting solution to ES-MS analysis of CPP digests in ladder sequencing is given by Rosnack and Stroh [14]. They tested continuous infusion of the digest mixture from a fused silica capillary reactor (volume 60–220 µl) held at 37°C in a water bath. The idea was to improve the interpretation of peptide ladders when the rate of cleavage differs by orders of magnitude for different residues as a consequence of the specific primary, secondary and tertiary structure. A continuous monitoring of the reaction in a flowing stream would then improve the situation and is also well suited for the ES application. Glucagon was tested, and a C-terminal sequence of 19 residues was obtained after about 200 min infusion starting from 1.9 nmol material (Tab. 1). Smith and Duffin [15] also used ES-MS to readout the peptide ladder resulting from CPP digestion, however, in a noncontinuous fashion. Recombinant interleukin 3 could be C-terminally sequenced for six residues using data from 10 and 50 min digestion time (Tab. 1). They furthermore demonstrated the effect of steric hindrance on ladder generation due to a polypeptide three-dimensional structure. The sequencing result is significantly improved upon reduction and carboxymethylation of cysteine residues. Superoxide dismutase yielded three C-terminal residues after this treatment, whereas only one was barely seen before breaking the disulfide interactions. Amounts consumed for analysis were in the range of 1–2 nmol [15].

Schär et al. [10] early tested identification of digested peptides via MALDI-MS for interpretation of amino acid sequences. They used a combination of CPY and CPB to determine eight consecutive residues from the C-terminus of the 34-residue human parathyroid hormone. The required amount was about 500 pmol to cover the necessary time-aliquots (Tab. 1). All measurements were made in the negative ion mode, and SA or 2,5-dihydroxybenzoic acid at 0.1 M in water were used as matrix [10]. The molar ratio of sample-to-matrix was 1:10,000. Digestion was carried out at 37°C, and eight aliquots were withdrawn at times varying from 1 to 120 min. The carboxypeptidase activity was quenched by mixing and drying the digest aliquot with the acidic matrix on the MALDI target. Schär et al. [10] combined data from trypsin digestion of the original peptide with ladder-sequencing data to be able to distinguish glutamine from lysine which have the same nominal mass. If the readout of the peptide ladder reveals a Gln/Lys residue and the tryptic digestion cleaves the peptide bond at this position, there is a good indication that this residue is lysine. On the other hand, if no cleavage occurs, the amino acid residue is most probably glutamine. Another approach to identification of lysine in relation to glutamine is derivatization to obtain a significant mass difference (below).

An important contribution was made by Patterson et al. [16] when they introduced the concentration-dependent digestion as a time- and material-saving alternative to the traditional time-dependent approach. Instead of collecting time-aliquots from a single digest, several parallel microdigestions are performed at different exopeptidase concentrations directly on a multiple sample-position stainless steel target [16]. Using CPY, they tested the strategy on 22 peptides of various charge, polarity and size. The largest peptide analyzed was 32 residues (ACTH fragment 7–38) and it was possible to readout 19 consecutive C-terminal residues (Tab. 1). The general recommendation is to prepare 10 microdigests, each containing 0.5 pmol of peptide, that are applied directly to the sample wells on the target plate. To 9 of these wells CPY is added and the 10th is used as control. The reaction is allowed to proceed at room temperature until the volume (1 µl) evaporates (approximately 10 min) and matrix (CHCA) is added followed by MALDI mass analysis [16]. The approach provides several advantages in relation to time-dependent digestions such as fast optimization of specific reaction conditions for a particular peptide and smaller consumption of both peptide and enzyme.

Even though CPY potentially releases all C-terminal amino acids from a peptide or a protein (including Pro), there are always practical limitations in terms of structure and suboptimal reaction conditions. To improve the generality, Thiede et al. [11] combined CPY (slow release of glycine and aspartic acid) and CPP (slow release of serine and glycine) to read up to 11 residues from the C-terminus using MALDI-MS, CHCA matrix and 500 pmol of starting polypeptide (Tab. 1). This length of C-terminal sequence was obtained for glucagon and a C-terminal myoglobin fragment generated by cleavage with CNBr. The latter indicates a strategy for exopeptidase digestion of large polypeptides and proteins. For these molecules the mass accuracy in MALDI-MS is often not sufficient, and for masses detected at a low signal-to-noise ratio, the accuracy is sometimes less than 0.1%. To be able to readout the peptide ladders properly and make the correct amino acid assignments, the protein can thus be cleaved into smaller fragments by endopeptidases or chemicals such as CNBr and analyzed in the same manner [11].

Bonetto et al. [17] also combined CPY and CPP to make the cleavage more efficient. Up to 20 C-terminal residues could be sequenced starting from approximately 500 pmol (Tab. 1). However, importantly, they tested derivatization of lysine and cysteine residues to improve the results. Peptides were treated with O-methylisourea to guanidinate lysine residues which are converted into homoarginine having a residue mass of 170.18 Da. In this manner lysine residues are easily assigned, and confusion with glutamine is avoided. Peptides that contain cystine as well as the common cysteine-modifications cysteic acid (after performic acid oxidation) or carboxymethylcysteine (after reduction and alkylation with iodoacetic acid) are refractory to proteolysis at those positions using CPY and CPP. Bonetto et al. [17] tested to convert cysteine to 4-thialaminine by (trimethylamino)ethylation with (2-bromoethyl)trimethylammonium bromide. The resulting peptide derivative allows successful carboxypeptidase degradation and identification of cysteine residues within the sequence. Furthermore, the derivatized lysine and cysteine side chains provide sites for protonation, which facilitates ionization in the MALDI-MS analysis which was carried out using the CHCA matrix [17].

A combination of peptide ladder sequencing (MALDI-MS) and collision-induced dissociation of peptides (ES-MS) for identification of amino acid exchanges in human hemoglobin variants was tested by Déon et al. [18]. A mixture of CPY and CPB was used, and up to five C-terminal residues were sequenced using 20–40 pmol of starting polypeptide and the CHCA matrix (Tab. 1). A combination of total mass determination, collision-induced dissociation and carboxypeptidase degradation was shown to be efficient for characterization of three hemoglobin variants [18].

To increase the amount of sequence information and provide a more comprehensive characterization, exopeptidase digestion from both the C- and the N-terminus has been tested on separate polypeptide aliquots. Caprioli and Fan [19] used CPY and LAP to digest peptides in the mass range 1300–2100 Da and FAB-MS to readout the ladders. After 2 h of CPY digestion of ribonuclease S peptide, eight residues had been removed from the C-terminus, and the corresponding truncated peptides could be identified in the FAB mass spectrum. Digestion of angiotensin III with LAP resulted in determination of four N-terminal residues, after which no further cleavage occurred, since LAP is unable to cleave the next prolylimide bond in the N-terminal sequence [19]. Using CPY and FAB-MS, Caprioli and Fan succeeded in C-terminally sequencing (Ile5) angiotensin II for the first three residues at a signal-to-noise ratio of approximately 3 : 1 using time-aliquots of only 660 fmol of peptide [19]. However, generally the method requires 0.5–5 nmol for sequence determination of up to eight residues (Tab. 1).

Woods et al. [20] combined MALDI-MS with on-plate time-dependent digestion using CPP and APM, and applied the technique to a nine-residue tumor-specific peptide antigen. After up to 10 min of CPP digestion starting with 500 fmol of peptide, the first five C-terminal residues could be determined, whereas APM digestion, employing another 500 fmol of starting material, resulted in three consecutive N-terminal residues (Tab. 1). Woods et al. tested the MALDI-MS sensitivity and showed that it was possible to detect the molecular ion of the synthetic nine-residue peptide after applying as little as 500 amol to the metal target using the CHCA matrix [20]. However, analysis of the native peptide recovered from cells (about 1×10^9) and purified via reverse-phase HPLC made it clear that the detection limit when contaminating peptides are also present is rather at the low femtomole level (10–30 fmol) [20]. Interestingly, Woods et al. found that the addition of a precipitating agent such as ammonium sulfate can significantly enhance the MALDI signal, which allows less material to be used in the ladder-sequencing procedure [20].

Finally, an interesting approach to C-terminal ladder sequencing combining enzymatic and chemical degradation has been presented by Thiede et al. [21]. In their method, CPP digestion is first applied followed by C-terminal chemical degradation based on an earlier protocol [22]. The same sample without transfer to a new Eppendorf tube can be used for both degradations, which required 30 pmol of peptide to reach at best eight C-terminal residues (Tab. 1). The idea is that the complementary action of enzyme and chemicals will result in extended sequence information and better assignment of residues. The latter is particularly true for lysine residues which become acetylated during the chemical treatment (activation by acetic anhydride, cf. Tab. 1), making them easy to distinguish from glutamine residues in the MALDI-MS readout (CHCA matrix) of the peptide ladder [21]. Interestingly, the peptide ladders can be formed by a single application of the chemical treatment; no repetitions were found to be necessary, which possibly is due to the extended reaction times (up to overnight).

3. Conclusion

Ladder sequencing with rapid one-step mass spectrometry readout of data is efficient and provides a strong complement to both N- and C-terminal sequence analysis of different types. The major advantage of ladder sequencing is the fast data acquisition and high sample throughput that is possible. Ladder sequencing allows multiple identifications and sequence assignment within minutes. The ladder-generating process requires a relatively short time, often less than 10–15 min if parallel concentration-dependent microdigestions are utilized. Furthermore, high sensitivity and low sample consumption are demonstrated for several ladder-sequencing protocols which are carried out at low picomole to femtomole polypeptide levels.

Acknowledgments
Work in the author's laboratory was supported by grants from the Swedish Medical Research Council (projects 13X-3532 and 13X-10832), the Swedish Cancer Society (project 1806), the Commission of the European Union (BIO4-CT97-2123), the Emil and Wera Cornell Foundation and the Magn. Bergvall Foundation.

References

1 Edman P (1950) Method for determination of the amino acid sequence in peptides. *Acta Chem Scand* 4: 283–293
2 Chait BT, Wang R, Beavis RC, Kent SBH (1993) Protein ladder sequencing. *Science* 262: 89–92

3 Bartlet-Jones M, Jeffery WA, Hansen HF, Pappin, DJC (1994) Peptide ladder sequencing by mass spectrometry using a novel, volatile degradation reagent. *Rapid Commun Mass Spectrom* 8: 737–742

4 Godovac-Zimmermann J, Krause I, Baranyi M, Fischer-Fruhholz S, Juszczak J, Erhardt G, Buchberger J, Klostermeyer H (1996) Isolation and rapid sequence characterization of two novel bovine β-lactoglobulins I and J. *J Prot Chem* 15: 743–750

5 Gu Q-M, Prestwich GD (1997) Efficient peptide ladder sequencing by MALDI-TOF mass spectrometry using allyl isothiocyanate. *J Peptide Res* 49: 484–491

6 Tsugita A, Takamoto K, Kamo M, Iwadate H (1992) C-terminal sequencing of protein. A novel partial acid hydrolysis and analysis by mass spectrometry. *Eur J Biochem* 206: 691–696

7 Takamoto K, Kamo M, Kubota K, Satake K, Tsugita A (1995) Carboxy-terminal degradation of peptides using perfluoroacyl anhydrides. A C-terminal sequencing method. *Eur J Biochem* 228: 362–372

8 Hayashi R, Moore S, Stein WH (1973) Carboxypeptidase from yeast. Large scale preparation and the application to COOH-terminal analysis of peptides and proteins. *J Biol Chem* 248: 2296–2302

9 Klarskov K, Breddam K, Roepstorff P (1989) C-terminal sequence determination of peptides degraded with carboxypeptidases of different specificities and analyzed by 252-Cf plasma desorption mass spectrometry. *Anal Biochem* 180: 28–37

10 Schär M, Börnsen KO, Gassman E (1991) Fast protein sequence determination with matrix-assisted laser desorption and ionization mass spectrometry. *Rapid Commun Mass Spectrom* 5: 319–326

11 Thiede B, Wittmann-Liebold B, Bienert M, Krause E (1995) MALDI-MS for C-terminal sequence determination of peptides and proteins degraded by carboxypeptidase Y and P. *FEBS Lett* 357: 65–69

12 Wang R, Cotter RJ, Meschia JF, Sisodia SS (1992) Determination of the cleavage site of the amyloid precursor protein by plasma desorption mass spectrometry. In: RH Angeletti (ed): *Techniques in protein chemistry III*. Academic Press, San Diego, 505–513

13 Wilm M, Mann, M (1996) Analytical properties of the nanoelectrospray ion source. *Anal Chem* 68: 1–8

14 Rosnack KJ, Stroh JG (1992) C-terminal sequencing of peptides using electrospray ionization mass spectrometry. *Rapid Commun Mass Spectrom* 6: 637–640

15 Smith CE, Duffin KL (1993) Carboxy-terminal protein sequence analysis using carboxypeptidase P and electrospray mass spectrometry. In: RH Angeletti (ed): *Techniques in protein chemistry IV*. Academic Press, San Diego, 463–470

16 Patterson DH, Tarr GE, Regnier FE, Martin SA (1995) C-terminal ladder sequencing via matrix-assisted laser desorption mass spectrometry coupled with carboxypeptidase Y time-dependent and concentration-dependent digestions. *Anal Chem* 67: 3971–3978

17 Bonetto V, Bergman A-C, Jörnvall H, Sillard R (1997) C-terminal sequence analysis of peptides and proteins using carboxypeptidases and mass spectrometry after derivatization of Lys and Cys residues. *Anal Chem* 69: 1315–1319

18 Déon C, Promé JC, Promé D, Francina A, Groff P, Kalmes G, Galactéros F, Wajcman H (1997) Combined mass spectrometric methods for the characterization of human hemoglobin variants localized within α T9 peptide: Identification of Hb Villeurbanne α89 (FG1) His → Tyr. *J Mass Spectrom* 32: 880–887

19 Caprioli RM, Fan T (1986) Peptide sequence analysis using exopeptidases with molecular analysis of the truncated polypeptides by mass spectrometry. *Anal Biochem* 154: 596–603

20 Woods AS, Huang AYC, Cotter RJ, Pasternack GR, Pardoll DM, Jaffee EM (1995) Simplified high-sensitivity sequencing of a major histocompatibility complex class I-associated immunoreactive peptide using matrix-assisted laser desorption/ionization mass spectrometry. *Anal Biochem* 226: 15–25

21 Thiede B, Salnikow J, Wittmann-Liebold B (1997) C-terminal ladder sequencing by an approach combining chemical degradation with analysis by matrix-assisted-laser-desorption ionization mass spectrometry. *Eur J Biochem* 244: 750–754

22 Schlack P, Kumpf W (1926) Über eine neue Methode zur Ermittelung der Konstitution von Peptiden. *Z Physiol Chem* 154: 125–170

Proteomics in Functional Genomics
ed. by P. Jollès and H. Jörnvall
© 2000 Birkhäuser Verlag Basel/Switzerland

New approaches for innovations in sensitive Edman sequence analysis by design of a wafer-based chip sequencer

Christian Wurzel and Brigitte Wittmann-Liebold*

WITA GmbH, Wittmann Institute of Technology and Analysis of Biomolecules, Warthestr. 21, D-14513 Teltow, Germany

Summary. In the last few years the development of new mass spectrometric techniques enabled fast and sensitive protein analysis by the introduction of mass spectrometry (MS) fingerprinting and MS/MS sequencing. For these methods mixtures of peptide fragments of the proteins can be employed, whereas the Edman degradation method requests purified peptides. On the other hand, Edman sequencing has the advantages that interpretation of the data is more simple, extended sequences can be derived, and reliable sequence information on unknown proteins is possible. Hence, Edman chemistry as an alternative technique to MS is still valuable. But higher sensitivity of the sequencers is needed in order to meet modern demands, e.g. in proteomics research. We designed a wafer-based chip sequencer for protein and peptide sequencing in the femto- to attomole range. The main advantage of our new design is the complete integration of dead volume free valves together with reactor and converter and volume-measuring loops within one wafer-based system. In this system aggressive chemicals and solvents for the Edman degradation can be delivered in sub-microliter amounts, which allows a considerable shortening of the degradation cycles. Further, we developed sensitive maintenance and tightness tests to prove a precise and reproducible delivery of the chemicals and the reduction of drying times as compared with conventional sequencers. Real parallel processing of several samples can easily be implemented. The system is designed to serve future needs in protein research.

Introduction

Proteomics, the characterization of the total cell protein set-up is a new approach to study cell processes and their dynamics at the molecular level [1]. In contrast to genomics, which provides the sequencing of all genes from a genome, the direct analysis of all proteins occurring in entire cells allows monitoring of individual cell states in development, regulation or disorders. Several important aspects in biology and medicine cannot be studied at all at the genome or messenger RNA (mRNA) level, e.g. the occurrence and functional role of posttranslational modifications, protein stability, enzyme activation or processing. But these studies can be approached by total protein comparisons of different cell situations. This is achieved by proteome analysis, which includes high-resolution two-dimensional (2D) gel electrophoresis in combination with analysis of the individual proteins by sensitive mass spectrometry (MS) and Edman sequencing.

---

* To whom the correspondence should be addressed.

Since Edman and Begg [2] developed sequencer for the stepwise amino acid sequencing of polypeptides, many improvements have been made in order to increase the reliability and sensitivity of the degradation to the low picomole range (see e.g. [3–7]). However, the sensitivity has to be further enhanced in order to enable the analysis of proteins and peptides, which are expressed in extremely low concentrations in cells and tissues. The best high-resolution 2D gels allow separation of up to 10,000 proteins [8] in complex mixtures from total cells or tissue extracts [9, 10]. New technical developments in MS facilitate measurements of peptides in mixtures, e.g. by MALDI-mass fingerprinting or by ESI-MS (electrospray ionization-mass spectrometry) techniques [11–13] and allow determination of partial sequences for identification of the protein spots by PSD-MALDI-MS or nanospray-ESI-MS/MS techniques [14–16]. For desalting and concentration of the protein digests, updated protocols were introduced, e.g. by attachment to reversed-phase beads and direct mass analysis of peptide mixtures. Recently, extended sequencing of proteins was also achieved by MALDI-MS of recombinant proteins, making use of the fast metastable decay immediately after laser irradiation [17]. On the other hand, disadvantages of these techniques are the time-consuming interpretation of the spectra and the fact that misleading data may be obtained. The interpretation of the data is complicated by the complexity of the ionization process, which leads to various sets of ions depending on the charge and fragmentation pattern. Limitations in reading the correct sequence arise also from the isobaric amino acids Leu/Ile and Gln/Lys, and the small difference of only 1 to 2 mass units, respectively, of Ile/Leu/Asp/Asn and Lys/Glu/Gln. Further, not all peptides "fly" at the ionization process, and even worse, this property is not predictable. Whereas identification and confirmation of known proteins by these techniques are mostly straightforward, the interpretation of the data for unknown or modified proteins is much more difficult. Hence, the alternative approach by direct amino acid sequence analysis employing Edman chemistry serves as an additional method that allows an unambiguous assignment of the protein sequences. Limiting factors are the amounts of purified proteins available and the sensitivity of the machines.

Present-day sequencers may be improved further by new and updated detection systems, such as capillary HPLC (high performance liquid chromatography). With ultraviolet (UV) absorbance, detection limits in the femtomolar range were reported [18]. Introduction of a faster measurement than gradient HPLC, which has the drawback of the time delay necessary for equilibration and washing of the column, would make a real parallel degradation of several samples possible. Another approach to the regular PITC (phenylisothiocyanate) chemistry is the separation by micellar capillary electrophoresis (CE) and the detection by thermooptical absorbance [19]. With this detection system, Dovichi and co-workers presented a miniaturized sequencer based on a teflon reactor and fused silica capillaries glued together. Disadvantages with this method are the absence of valves to prevent cross-contaminations and the exposure to oxygen [20]. Wu and Dovichi [21] showed separation of FTH amino acids by CE and detection limits in the zeptomolar range by laser-induced fluorescence detection. With the reintroduction of the double coupling method [22], they tried to overcome low repetitive yields of the fluorescent isothiocyanate derivatives [23]. A drawback of CE separation is that only a very small portion of the cleaved and converted amino acid is injected into the separation system, whereby the profit of the sensitive detection gets lost. MS generally offers high sensitive detection, and therefore several such approaches to detect the released amino acids from Edman degradations were proposed (see [24], or for a short review [25]). Unfortunately the PTH amino acids show only poor ionization rates, and none of the methods reported a breakthrough. Recently, Zhou et al. showed [25] higher ionization rates and subfemtomole detection limits of some PTH amino acids on ESI-MS by adding lithium triflate to the buffer, which might be promising. Aebersold and co-workers tried to find isothiocyanate derivatives, which have a high coupling yield and show better ionization rates in MS. With 4-(3-pyridinylmethylaminocarboxypropyl)-PTH amino acids they showed low femtomol detection limits [26]. Unfortunately the PITC derivative is not commercially available.

1. Limiting factors of present sequencers

In present-day sequence automates the amounts of reagents and solvents delivered are in the range of 5 – 100 µl. Although reagents and solvents used in Edman degradation are highly purified, they still contain traces of impurities, aldehydes, oxygen and water. Further, the delivered amounts lead to a washout of the sample, high release of by-products due to side reactions of Edman degradation and an enhanced noise level. Due to the amount of remaining contamination in the reagents and solvents, a partial blockage of the protein may occur, decreasing initial and repetitive yields. These impurities can only be reduced significantly by decreasing the amounts of delivered reagents and solvents. The limits become more and more severe with the reduction of sample amounts applied. Other reasons for side reactions are memory effects. Reagents or impurities stick to the surfaces. This effect is increased by unevenly cut channels of the liquid flow path with undercuts, hidden edges or rest volumes in the ferrules. Additionally, regarding the memory effect in teflon, the relatively long tubings as connecting lines show oxygen penetration. Further, relatively large volumes within the valves and within the reaction chamber and converter require extensive washings and prolonged drying times, leading to a washout of the sample and high consumption of solvents. Today there is a misfit between the sample amounts available, for instance in proteome analysis, and the size (inner volume and surface) of the delivery channels on the one hand, and of the reaction chamber and reagent and solvent consumption in the machines on the other. In the past 2 decades no significant reduction in the respective sizes and reagent consumption was achieved, although the applied sample amounts were reduced to the low picomole range. Further, an increase in the sensitivity of the detection of the released amino acid derivatives alone would not yield a drastic change in current sequencer technology. The other part, namely the liquid handling system of Edman chemistry and the interface between the chemistry conducting part and the detector device, must also be accommodated to minute protein samples.

2. Liquid handling with microsystems technology

A possibility to miniaturize the reactor flow channels or the valves of the sequencer technology would be to manufacture a sequencer in microsystems technology. This technology offers a wide variety of different structuring processes, e.g. the LIGA technique [27] or etching in silicon or glass. The development of micro-machines such as motors or gears, mechanical and chemical sensors, liquid-handling devices such as pumps, valves and mixers, and the adaptation of the structuring processes to the dimensions of fluid systems or microreactors are the main focus in micro-reaction technology [28, 29]. In comparison to conventional reactors or heat exchangers, the micro-reaction technology has the advantage of very high surface-to-volume ratio. Although the concept of a micro total analytical systems [30] was introduced relatively early, these and other systems remained confined to the laboratory. Thus far, applications of microsystems technology are limited to single components such as acceleration sensors, pressure sensors or printer heads with ink jet technology, but the market and the number of applications are still growing. A separation technique which can easily be adapted to the microscale is electrophoretic separation [31]. Here microsystems technology has the advantage that injection of the sample and appropriate detection cells can be integrated. The transport of liquids, mixing and injection within the chip are governed by switching different potentials between several in- and outlets [32]. Disadvantages of this technique are that the liquids must have conducting properties and no hermetic disclosure of a reagent or the sample can be achieved. In fact, the opposite is the case: the liquids must stay in contact, or the electroosmotic flow cannot take place. Such techniques are applicable to e.g. pre- or postseparation derivatizations, polymerase chain reaction (PCR) devices or an automated sample injection to ESI-MS [33].

3. Design of the chip derivatization unit

A goal of the newly designed micro-reaction system on the one hand is the miniaturization of volumes and surfaces. Another is the integration of a

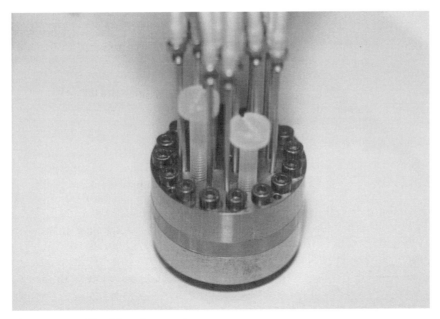

Figure 1. Assembly of the micro-reaction system.

multiple valve system together with the reactor and converter in one device. The external connecting lines between these units are then eliminated. The construction consists of a multilayer arrangement within a housing made from aluminium. Figure 1 shows a picture of the mounted system in a prototype version.

The valves free of dead volume, the delivery channels, the reaction chambers, additional internal loops and the connections to the reagent reservoirs are fully integrated within the chip. A core part of the system is a wafer, a central disk of 3–5 cm and a thickness of 2 mm, consisting of photostructured glass, perfluorated polymer or sintered ceramics. Long-term tests ongoing in our laboratory will show which material has the best chemical resistance. In special manufacturing processes, a structure with holes through the wafer as well as channels at the surface is built. The whole arrangement on the wafer is designed for protein microsequencing. Eighteen lines for the inlet of reagents, solvents and inert gas and outlets to waste and to the detection system are integrated. Each line is closed, dead-volume-free by a membrane valve, all of which can be operated individually. The delivery lines of 100 μm diameter connect the inlets to the reaction or conversion chamber within the chip system. The delivery of the reagents towards the channel is done by a slight overpressure of nitrogen or argon within the reservoir bottles. In the channels the delivered liquid droplets are transported to the reactor or converter by application of nitrogen or argon pressure to the channel through the end valve, an inert gas inlet. The

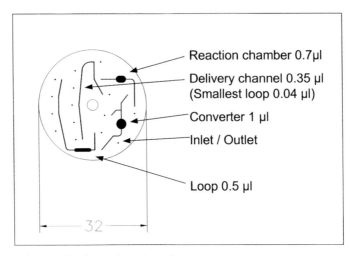

Figure 2. Scheme of the flow path on the wafer.

amount of liquid delivered is either time-controlled or exactly measured by volumetric loops which are provided within the wafer as well. A scheme of the micro-reaction system designed for Edman degradation is shown in Figure 2. Indicated with arrows are the inner volumes of loops and reaction chambers. Every inlet and outlet line is closed by an integrated membrane valve.

Figure 3 shows another design in a prototype version of the chip system made of photostructured glass. The upper surface of this wafer has photo-structured channels, one each as delivery channel for the reactor and for the converter, and the third for collection of the degraded amino acid deriva-tives towards the detection system. Close to the respective channels short holes through the wafer are arranged which serve as inner connections be-tween the upper side of the wafer and the reservoir flasks arranged below.

In addition to valves for the delivery of reagents and inert gas which guide the fluids to the reaction chamber or to the converter, respectively, the wafer contains some additional valves for measurements of solvents and for collection of waste. Each of the channels can be closed hermetically or washed independently. The connections from the reagent bottles and the inert gas lines to the wafer are made from glass capillaries or TEFZEL tubings connected to the bottom side of the wafer. The upper structured side of the wafer is completely covered by a flexible membrane. Short channels lead from the inlet holes to the delivery channels also closed by the membrane. The valves can be operated individually by lifting the membrane dome-like by applying vacuum. They are closed by applying nitrogen pressure. Thus holes in the upper part of the aluminium housing are connected to 3/2-way pneumatic valves. This device realises a dead-volume-free membrane valve function between inlet and delivery channel.

Table 1. Carbohydrate-amino acid linkage types found in glycoproteins (the three most often occurring linkages are indicted in boldface)

Class	Linkage	Class	Linkage
N-linked		*Other O-linked*	
	GlcNAc → Asn		Glc → Tyr
	GalNAc → Asn		Gal → OH-Lys
	Glc → Asn		L-Ara → OH-Lys
	L-Rha → Asn		L-Ara*f* → OH-Pro
O-linked			Gal → OH-Pro
			Gal → OH-His
	GalNAc → Ser/Thr		GlcA → OH-Trp
	GlcNAc → Ser/Thr		GlcA → OH-Phe
	Man → Ser/Thr		GlcA → OH-Ser
	L-Fuc → Ser/Thr		
	Gal → Ser/Thr	*S-linked*	
	Glc → Ser		Gal → Cys
	Xyl → Ser		Glc → Cys
ADP ribosylation		*C-linked*	
	ADP-Rib → N^{ξ}-Arg		Man → Trp
	ADP-Rib → N^{δ}-Asn		
	ADP-Rib → N^{1}-His	*Glycation*	
	ADP-Rib → *O-C(O)*Glu		Glc → Lys
	ADP-Rib → *S*-Cys		Rib → Lys
Phosphoglycosylation		*Amide Bond*	
	GlcNAc1-PO$_4$ → Ser		GlcA/GalA(6 → N^{α})Lys
	Man1-PO$_4$ → Ser		GlcA/GalA(6 → N^{α})Thr/Ser
	Xyl1-PO$_4$ → Ser/Thr		GlcA/GalA(6 → N^{α})Ala
			MurNAc(3 → N^{α})Ala
Glypiation			
	Protein-C(O)-NH-(CH$_2$)$_2$-PO$_4$ → 6Man (GPI)		

glycans can be built up in glycoproteins due to elongation and branching. A glycoprotein often contains more than one glycosylation site, and on a single glycoprotein the glycan structures at each of the sites can comprise a family of glycans. This phenomenon is known as microheterogeneity. However, the ensemble of glycoforms has a fairly constant composition for a given protein from a specific source. Nowadays, commercial kits are available for sensitive detection of glycosylated proteins.

The occurrence of glycopeptides in nature, for instance in urine, is predominantly due to degradation of glycoproteins. However, in insects, some of the antibacterial proline-rich peptides (2–3 kDa) are O-glycosylated [11]. Several antibiotics (e.g. vancomycin, teicoplanin, ristocetin, eremomycin), isolated from microorganisms, have proved to be complex glycopeptides containing rare amino sugars [12].

The term "glycopeptide" should not be confused with peptidoglycan. Peptidoglycan, found particularly in bacterial cell walls, has a polysaccharide backbone of alternating GlcNAc and *N*-acetylmuramic acid

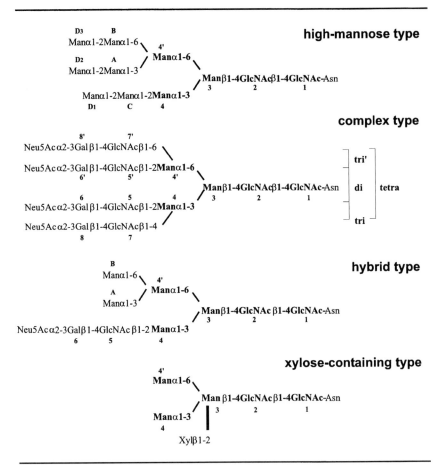

Figure 1. Examples of the four types of N-linked glycans of glycoproteins. The common pentasaccharide core is shown in boldface. For the complex type the di-, tri-, tri′- and tetra-antennary subtypes are indicated. The standard notation of the monosaccharide residues has been included.

(MurNAc). Some of the MurNAc residues contain a tetrapeptide side chain, and the polysaccharide is cross-linked with peptide bridges [10].

2. Glycoprotein analysis

In higher animals most of the proteins essential for life, like immunoglo-bulins, enzymes, cell-membrane receptors and hormones, are in fact glyco-proteins, and their functioning greatly depends on the glycan part. To in-vestigate the metabolism and molecular biology of these biomolecules, knowledge of their precise structures and in particular of the glycan struc-

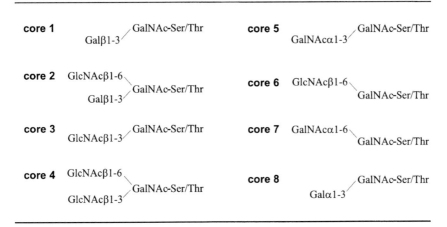

Figure 2. Core structures of mucin-type O-linked glycans of glycoproteins. The core structures can be extended with Gal, GalNAc, GlcNAc, Fuc and sialic acid residues.

tures is indispensable. Furthermore, the impressive increase of the number of biotechnologically produced glycoprotein-therapeutics demands accurate, fast and reliable analysis methods for determination of composition and structure [13–16].

Structural analysis of glycoproteins involves several levels of detail: (i) the amino acid sequence of the protein; (ii) the number, nature and position of glycosylated amino acids; (iii) the identity and quantity of monosaccharide residues present in the protein; (iv) the structural class(es) of oligosaccharides present in the protein; (v) the nature and structural heterogeneity of the glycans attached at specific sites in the protein; (vi) the monosaccharide sequence, branching and glycosidic linkage types of each glycan; and (vii) the secondary and spatial structure of the glycoprotein, in particular, the protein and the carbohydrate part, and their mutual influence. For these studies it is essential that adequate starting material is available, which means that much care has to be bestowed on the isolation and purification of the glycoprotein. A glycoprotein that is apparently homogeneous with respect to the protein backbone may nevertheless comprise a collection of glycoforms.

The structural characterization of glycoproteins is generally considered as a relatively difficult problem, in particular because it requires the use of highly sophisticated instrumentation. In view of the diversity and complexity of the glycans, their primary structure analysis has still not reached the level of routine analysis and remains a specialized and laborious endeavour depending on the combined use of several physical, chemical and biochemical techniques [17].

The discussion of the analysis of glycoproteins will be divided into three topics, namely, (i) the protein/peptide part, (ii) the glycan part and (iii) the

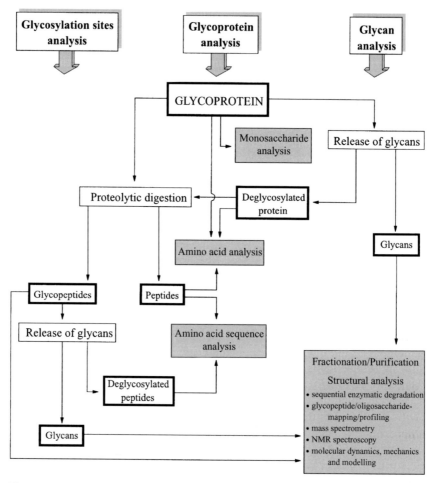

Figure 3. A general strategy for the analysis of glycoproteins/peptides.

determination of the glycosylation sites. Moreover, it will be focused on the preparation and analysis of glycopeptides. A general strategy for glyco-protein glycan analysis is presented in Figure 3.

2.1. Protein/peptide analysis

The presence of carbohydrate chains linked to the protein is no obstacle for the determination of the amino acid composition. Amino acid analysis is generally performed, after acid hydrolysis, on an Amino Acid Analyzer using ion-exchange chromatography. Absorbance is detected by post-column treatment with ninhydrin. However, it has to be noted that amino-sugars, if present in large amounts, can give rise to peaks in the chromatogram coeluting with certain amino acids.

For the determination of the complete amino acid sequence, the protein is usually partially hydrolyzed into peptides by proteolytic enzymes [this book]. In this case, covalently linked carbohydrates, in particular when adjacently clustered at specific regions of the polypeptide backbone (e.g. in mucins), can seriously hamper the action of the proteases. Preceding excision of the carbohydrate chains without destroying the protein or peptide may be achieved chemically (e.g. O-glycans by mild alkaline borohydride treatment) and/or enzymatically (e.g. N-glycans by PNGase) (*vide infra*). Also, the use of anhydrous trifluoromethanesulfonic acid (TFMSA) has been found to be successful [18]. In this case, N- and O-glycans are cleaved nonselectively, leaving the primary structure of the protein intact for sequence analysis, but the carbohydrate chains are destroyed. The best results to deglycosylate heavily O-glycosylated proteins have been obtained by a combined approach using mild TFMSA treatment, followed by periodate oxidation and alkaline treatment [19].

2.2 Glycan analysis

In order to get an overview of the carbohydrate portion of a protein, the intact glycoprotein is usually first subjected to a monosaccharide analysis. The determination of the monosaccharide composition and carbohydrate content is currently established after methanolysis and analysis of volatile sugar derivatives by gas chromatography(-mass spectrometry) (GC-MS) [20]. Recently, high-pH anion-exchange chromatography with pulsed amperometric detection (HPAEC-PAD) has been introduced for monosaccharide analysis using underivatized monosaccharides obtained by trifluoroacetic acid (TFA) or HCl hydrolysis [21], but different hydrolysis conditions are needed for quantitative determination of different types of monosaccharides.

Once the monosaccharide compostion of the glycosylation is known, putative structural classes of glycans present on the glycoprotein can be inferred. The detection of GlcNAc, Man, Fuc and Neu5Ac indicates the presence of N-glycans, whereas detection of relatively large amounts of GalNAc indicates possibly O-glycans. These initial clues are beneficial for further tackling the structural analysis of the glycans.

Since glycoproteins contain mostly multiple glycosylation sites, and the oligosaccharide chains usually exhibit considerable structural heterogeneity, it is hardly possible to analyze complete glycan structures at the level of the intact glycoprotein. The exhaustive nonselective proteolytic cleavage of the protein and subsequent analysis of the isolated glycopeptides is a good approach to establish the glycan structures [22]. In this case, it is important that the number of glycosylation sites comprised in the structure is known.

An alternative practical approach is to liberate the carbohydrate chains, then to separate the oligosaccharide pool from the deglycosylated protein, and to fractionate the oligosaccharides. Glycans can be chemically and/or enzymatically released from the protein backbone (Tab. 2). Hydrazinolysis [23] is frequently applied to liberate N-glycans, and depending on graded reaction conditions, sequential release of O-glycans followed by N-glycans may be possible [24]. Other chemical methods are reductive deamination [25] and trifluoroacetolysis [26]. It must be noted that chemical treatments may lead to some degradation of the carbohydrate portion and (partially) destroy the polypeptide backbone. The release of O-glycans can be realized with alkaline borohydride treatment (β-elimination reaction) [27].

In the enzymatic approach two types of enzymes are currently used, namely, the endo-β-N-acetylglucosaminidases (endoglycosidases) and the peptide-N^4-(N-acetyl-β-D-glucosaminyl) asparagine amidases (PNGases) to liberate N-glycans. For both types of enzymes, the efficiency of release depends on their specificity and the type of oligosaccharide structures [28]. For application to glycopeptides, introduction of restrictions by the length of the peptide backbone must be taken into account (Tab. 2). The enzymatic release of O-glycans by endo-N-galactosaminidases (O-glycanases) is very restricted because of their limited specificities and therefore rarely used [29]. Occasionally, it may be necessary first to denature the glycoprotein to make possible complete oligosaccharide release. Care should always be

Table 2. Chemical and enzymatic release of glycans from glycoproteins/peptides

Chemical Procedure	Specificity/target	Comments
TFMSA treatment (trifluoromethane sulfonic acid)	N- and O- glycans	– glycans are destroyed
Hydrazinolysis (eventually with concomitant re N-acetylation and reduction)	N- and O- glycans (depending on conditions)	– protein is destroyed – loss of N- and O-acyl groups – partial loss of sulfate and phosphate
Reductive deamination (nitrous acid hydrolysis)	GPI anchor	– cleavage between GlcNH$_2$ and inositol
Alkaline borohydride treatment (β-elimination reaction) (often with concomitant reduction)	O-glycans (N-glycans to a certain extent)	– peeling reactions – loss of O-acetyl groups – not active on N-terminal Thr/Ser glycan – possible destruction of protein
Trifluoroacetolysis (mixture TFA:TFAA 1:100)	glycosylamine linkage is cleaved (peptide bonds)	– oligosaccharides are released as TFA derivatives – Fuc attached to GlcNAc1 is destroyed – low recovery

Table 2 (continued)

Enzymatic Enzyme	Abbreviation	Source	Specificity
Endoglycosidase D (EC 3.2.1.96) Endo-β-N-acetyl glucosaminidase D	Endo D	*Diplococcus pneumoniae*	– N-glycans – cleavage between GlcNAc residues of chitobiose-Asn – high-mannose (Man$_5$GlcNAc$_2$) types – not active on intact complex and hybrid types
Endoglycosidase F (EC 3.2.1.96) (mixture of endo-glycosidases F$_1$, F$_2$, F$_3$) Endo-β-N-acetyl glucosaminidase F	Endo F	*Flavo-bacterium meningo-septicum*	– N-glycans – cleavage between GlcNAc residues of chitobiose-Asn – high-mannose, diantennary complex and hybrid types
Endoglycosidase H (EC 3.2.1.96) Endo-β-N-acetyl-glucosaminidase H	Endo H	*Streptomyces griseus* (*S. plicatus*) (recombinant in *E. coli/S. lividans*)	– N-glycans – cleavage between GlcNAc residues of chitobiose-Asn – high-mannose (not sulfated) types – some hybrid types – not active on complex types
Peptide-N-Glycosidase F (EC 3.2.2.18) (EC 3.5.1.52) peptide-N$_4$-(N-acetyl-β-D-glucosaminyl) asparagine amidase Glycopeptidase F N-glycosidase N-glycanase	PNGase F	*Flavo-bacterium meningo-septicum* (recombinant in *E. coli*)	– N-glycans – cleavage between GlcNAc-Asn (Asn converted into Asp) – high-mannose, complex and hybrid types – not from a single or N/C-terminal Asn – hardly active when core Fucα1-3 present – substituents remain
Peptide-N-Glycosidase A (EC 3.5.1.52) peptide-N$_4$-(N-acetyl-β-D-glucosaminyl) asparagine amidase Glycopeptidase A N-glycosidase	PNGase A	Almond emulsine	– N-glycans – cleavage between GlcNAc-Asn – high-mannose, complex and hybrid types – not from a single or N/C terminal Asn – peptide backbone restrictions
Endo-α-N-acetyl-galactosaminidase (EC 3.2.1.97) (EC 3.2.1.110) O-Glycosidase O-glycan peptide hydrolase O-Glycanase	–	*Streptococcus* (*Diplococcus*) *pneumoniae* (*Alcaligenes* sp.)	– O-glycans – only cleavage of disaccharide Galβ1-3GalNAcα1- from Thr/Ser

taken to avoid glycan modifications, including loss of sialic acid residues, Fuc, and noncarbohydrate substituents.

A convenient strategy for the study of glycans of N,O-glycoproteins is based on cleavage of the N-glycans with PNGase-F, followed by alkaline borohydride release of the O-glycans from the remaining purified O-glyco-protein [30]. In case of minimal amounts, labeling [e.g. with 2-amino-benzamide (2-AB), aminonapthalenesulfonic acid (ANTS) or tritium] of the released glycans is useful to enable their detection in subsequent procedures.

2.3 Glycosylation site analysis

A suitable approach to characterize the glycosylation sites in a glycoprotein consists of degrading the protein backbone. Selective mild proteolytic digestion followed by fractionation of the formed glycopeptides is applied for this purpose. The location of the glycosylation sites in the polypeptide chain is determined by analysis of the amino acid sequence around each glycan attachment site of the glycopeptide and comparison of this sequence with the known amino acid sequence of the protein. Identification of the glycan structure and of the peptide chain are performed before and after removal of the glycan.

3. Glycopeptide analysis

In the framework of this review glycopeptides are defined as glycoconju-gates that contain oligosaccharides still attached to a portion of the original peptide sequence (oligopeptide) of the glycoprotein. Preparation of glyco-peptides from glycoproteins represents an important step in the determina-tion of the glycan structures, but moreover it is essential for the determina-tion of the glycosylation sites and of the site-specific (micro)heterogeneity. For the study of cell surface glycosylation, the preparation and analysis of glycopeptides is a general procedure [31]. Glycopeptides can chemically be prepared by cyanogen bromide (CNBr) cleavage of glycoproteins at the carboxylic side of Met [32]. Sometimes this treatment is necessary prior to proteolytic digestion of glycoproteins, which have a tendency to form aggregates [33].

3.1 Proteolytic digestion

Proteolytic digestion may be performed either on native or on denatured glycoproteins and should be optimized for each glycoprotein. The size and type of glycopeptides formed are dependent on the amino acid sequence of the protein, the glycan structures present in the protein and the specificity of the proteases being used. Some frequently used proteolytic enzymes are listed in Table 3.

Denaturation of proteins is often achieved by heat treatment and/or reduction of cystine bridges followed by carboxymethylation of the free SH groups [31].

Nonselective proteolysis with pronase, which is a mixture of endopeptidases, is often used to obtain glycans linked to one amino acid or glycopeptides with a very short (two or three amino acids) peptide chain. Proteases with restricted selectivity, such as trypsin or chymotrypsin, are used when a longer peptide chain is required to assess the different glycan structures at each of the glycosylation sites. Today, endoproteinase Glu-C hydrolysis is frequently used for glycosylation site analysis. However, care must be taken because often incomplete and aspecific proteolytic cleavages give rise to complex mixtures of peptides and glycopeptides, including the possibility that one glycosylation site can be represented by more than one glycopeptide. In O- and N,O-glycoproteins, multiple O-glycosylation sites may occur that are clustered in certain regions of the polypeptide backbone, thereby impeding the preparation of glycopeptides containing a single O-glycosylation site. Consequently, in heavily O-glycosylated proteins (e.g. mucins), the assignment of a particular glycan to a specific amino acid is difficult to achieve.

Table 3. Some proteolytic enzymes frequently used in glycoprotein studies

Enzyme	Source	pH optimum	Specificity	Comments
Carboxypeptidase A (EC 3.4.17.1) Carboxypoly-peptidase	bovine pancreas	7.0–8.0	– successive cleavage from C-terminus – slow for Gly, Asp, Glu, Cys	Zn metallo-protease – does not act at Arg, Lys, Pro(OH)
Carboxypeptidase B (EC 3.4.17.2) Protaminase	porcine/hog pancreas	7.0–9.0	– successive cleavage from C-terminus of basic amino acids (Lys, Arg)	Zn metallo-protease
Chymotrypsin (α) (EC 3.4.21.1)	porcine/ bovine pancreas	7.5–8.5	– C-terminal bonds of Tyr, Phe, Trp – slow for Met, Leu, Ala, Asp, Glu	Serine endopeptidase
Endoproteinase Arg-C (EC 3.4.21.40)	Mouse sub-maxillaris glands *Clostridium histolyticum*	8.0–8.5	– carboxylic side of Arg	Serine protease
Endoproteinase Asp-N (EC 3.4.24.33)	*Pseudomonas fragi* (mutant) *Clostridium histolyticum*	7.0–8.0	– amino side of Asp (Glu) and cysteic acid	Metalloprotease

Table 3 (continued)

Enzyme	Source	pH optimum	Specificity	Comments
Endoproteinase Glu-C (EC 3.4.21.19) V-8 protease proteinase V8	*Staphylococ-cus aureus* V8	4.0 and 7.8	– carboxylic side of Glu (and Asp)	Serine protease
Endoproteinase Lys-C (EC 3.4.99.30) (EC 3.4.21.50)	*Lysobacter enzymogenes*	8.5–8.8	– carboxylic side of Lys	Serine protease
Papain (EC 3.4.22.2)	*Carica papaya*	6.0–7.0	– Arg, Lys, Glu, His, Gly, Tyr – total hydrolysis on prolonged incubation – does not act at acidic residues	Cysteine endopeptidase (thiol protease)
Pepsin (EC 3.4.23.1)	pig gastric mucosa	2.0–4.0	– preferentially carboxylic side of Phe, Met, Leu, Trp – Tyr-X and X-Val/Ala/Gly are relatively resistant	Aspartic endopeptidase (carboxyl protease) (acid protease)
Pronase (EC 3.4.24.4) Actinase E	*Streptomyces griseus*	7.5–8.0	– total hydrolysis	Mixture of several unspecific endo- and exoproteases
Subtilisin (EC 3.4.21.14) (EC 3.4.21.62) Alcalase/Nargase	*Bacillus subtilis* *B. licheniformis*	7.0–8.0	– total hydrolysis – preferentially Asp, Glu, Ala, Gly, Val	Serine endopeptidase
Thermolysin (EC 3.4.24.4) (EC 3.4.24.27)	*Bacillus thermo-proteolyticus*	7.0–9.0	– low cleavage specificity Ile, Leu, Met, Phe, Trp, Val, Ala	Zn metalloendo-peptidase thermostable 4–80°C
Trypsin (EC 3.4.21.4)	bovine/hog pancreas	7.5–8.5	– carboxylic side of Arg, Lys	Serine endopeptidase

3.2 Isolation and purification of glycopeptides

A large variety of high-resolution separation techniques are available today for the isolation, fractionation and purification of glycopeptides (and oligosaccharides), as summarized in Table 4. A combination of these methods is advisable, and often necessary, to obtain homogeneous compounds. After nonspecific proteolytic treatment of the glycoprotein, the first step is the separation of the glycopeptides from the bulk of small peptides, amino acids and salts. This can be achieved by size exclusion (gel filtration) chromatography. To facilitate the isolation and analysis procedures,

Table 4. Separation techniques frequently used for glycoproteins/peptides and their glycans

Separation method	Abbreviation	Separation parameter	Target compounds (analysis)
Chromatography			
gel filtration size exclusion gel permeation	GFC SEC GPC	molecular size and shape (hydro- dynamic volume)	– glycoproteins – glycopeptides – oligosaccharides (size profile)
ion-exchange anion-exchange cation-exchange (HPLC)	IEC AEC CEC	net charge	– glycoproteins – glycopeptides – sialyloligosaccharides (charge profile)
high-pH anion- exchange chromatography	HPAEC	charge/size (oxy-anions) hydrogen bonds dipole interaction	– (glycopeptides) – monosaccharides – oligosaccharides (oligosaccharide mapping)
normal-phase high- performance liquid chromatography	NP-HPLC	hydrogen bonding and/or partition (hydrophilic inter- action) (size/polarity)	– neutral (acidic) oligosaccharides (high-mannose type N-glycans)
reversed-phase high- performance liquid chromatography	RP-HPLC	hydrophobic inter- action	– glycopeptides – (permethylated) oligosaccharides (sialic acid analysis)
affinity chromatography on immobilized lectins	AC	biospecific inter- action	– (glycoproteins) – glycopeptides – oligosaccharides
Electrophoresis			
preparative paper electrophoresis (high/low voltage)	PPE	charge	– (glyco)peptides – oligosaccharides
(sodium dodecyl- sulfate)-polyacrylamide gel electrophoresis	(SDS)/ PAGE FACE™	molecular mass/ charge	– glycoproteins – deglycosylated proteins (N-glycan analysis)
isoelectric focusing (polyacrylamide gel)	IEF	charge/pI (pH gradient)	– glycoproteins
capillary zone electrophoresis	CZE	mass/charge ratio (pH/ionic strength)	– glycopeptides – charged oligo- saccharides
capillary gel electrophoresis (polyacrylamide)	CGE	molecular size charge/mass ratio constant	– glycoproteins – glycopeptides – oligosaccharides (glycoform mapping)
affinity capillary electrophoresis	ACE	charge/biomolecular interaction	– lectin-sugar binding

N-terminal amino acids in the glycopeptides can be labeled by N-(^{3}H)- or N-(^{14}C)-acetylation or N-dansylation, and elutions be performed with volatile buffers.

Reversed-phase high-performance liquid chromatography (RP-HPLC) is most commonly used for glycopeptide purification [34]. The separation is based on the hydrophobicity of the peptide part. Hydrophobicity can be

Table 5. Lectins frequently used for carbohydrate identification and separation in glycoprotein studies

Lectin	Abbreviation	Source	Specificity
Agaricus bisporus	ABA	mushroom	Galβ1-3GalNAc
Anguilla anguilla	AAA	fresh water eel	Fucα, Me-C2/C3 Fuc
Arachis hypogaea	PNA	peanut	Galβ, Galβ1-3GalNAc, Galβ1-4Glc
Canavalia ensiformis	Con A	Jack bean	Manα, Glcα, GlcNAcα, Me-αMan, branched mannoses
Datura stramonium	DSL	Jimson weed, thorn apple	GlcNAcβ, GlcNAcβ 1-4GlcNAc oligomers, Galβ1-4GlcNAc, N-acetyllactosamine repeats
Dolichos bifloris	DBA	horse gram	terminal GalNAcα, GalNAcα1-3GalNAc
Erythrina cristagalli	ECA	coral tree	Galα/β, GalNAcα/β, Galβ1-4GlcNAc
Galanthus nivalis	GNA	snowdrop bulb	Manα, Manα1-3Man, terminal Manα1-3
Glycine max	SBA	soybean	terminal GalNAcα/β, GalNAcα1-3Gal
Helix pomatia	HPA	Roman snail	GalNAcα/β
Lens culinaris	LcH	lentil	Manα, Glcα, GlcNAcα, branched mannoses with αFuc, Fucα1-6coreGlcNAc
Lotus tetragonolobus	LTA	asparagus pea	Fucα, Fucα1-2-Galβ1-4-[Fucα1-3]GlcNAc
Maackia amurensis	MAA	–	Sialic acid, Neu5Acα2-3Gal, (Galβ1-4Glc)
Maclura pomifera	MPA	osage orange	Galα, GalNAcα
Narcissus pseudonarcissus	NPA	daffodil	terminal and internal Manα
Pisum sativum	PSA	garden pea	Manα, Glcα
Sambucus nigra	SNA	elderberry	Galβ, Sialic acid, Neu5Acα2-6Gal(NAc)
Triticum vulgaris	WGA	wheat germ	GlcNAcβ, sialic acid, GlcNAcβ1-4GlcNAc
Ulex europaeus	UEA-1	gorse, furze	Fucα, Fucα1-2Galβ1-4GlcNAc

increased by modification of the amino-terminus with *tert*-butyloxycar-bonyl tyrosine. Since peptide sequences surrounding each glycosylation site, in particular for N-glycosylation, are usually quite different, RP-HPLC often resolves glycopeptides from multiple glycosylation sites derived from the same glycoprotein. The glycosylation-site-specific glycopeptide mixture can be further fractionated on the basis of carbohydrates. Separation based simultaneously on peptide and oligosaccharide structures can also be accomplished with RP-HPLC [35]. Anion-exchange (AE) HPLC has been used to separate glycopeptides, but the method is highly dependent on the structure of the peptide portion. Elutions are usually monitored by ultraviolet (UV) absorbance.

For the detection of the glycopeptide fractions among the many peptide fractions in a chromatogram, different strategies are applied: (i) mono-saccharide analysis or thin-layer chromatography (TLC)-spottest on each collected peak; (ii) amino acid sequence analysis, which usually indicates the presence of amino acid-linked oligosaccharides, on each collected peak; (iii) tandem MS/MS; (iv) enzymatic cleavage of the glycans in the (glyco)peptide pool causing a shift in the HPLC elution time; and (v) lectins conjugated to antibodies or enzymes to detect the presence of carbohydrates in collected fractions [36]. The properties of lectins to bind specifically to a certain sugar and/or sugar sequence in oligosaccharides and glycoconjugates (Tab. 5) are applied in affinity chromatography to fractionate and purify glycopeptides by use of immobilized lectin columns [37, 38], and more recently, by affinity capillary electrophoresis for micro-scale analyses [39]. Improvements in chromatographic techniques have been major factors in carbohydrate-research advances.

4. Structural analysis of glycans

The two main methods to determine the primary structure of carbohydrate chains are mass spectrometry (MS) and nuclear magnetic resonance spectroscopy (NMR). They are applied to purified oligosaccharides and glycopeptides, but when mixtures are not too complicated, sometimes the analysis of mixtures is possible. In some cases, additional evidence from chemical and/or enzymatic analytical methods is still necessary. This holds in particular for the unambiguous characterization of novel structures. The information from more than one analytical method (Tab. 6) is needed to arrive at conclusive evidence on the identity of a compound.

4.1 Chemical and enzymatic methods

It is evident that GC-MS monosaccharide composition analysis of the purified glycopeptides or oligosaccharides already provides information

Table 6. Methods of obtaining information about specific carbohydrate features of glycoprotein glycans

Information	Methods
Carbohydrate content, composition, D/L-configuration	colorimetric determinations, GLC-monosaccharide analysis, GLC-absolute configuration determination, NMR spectroscopy
Molecular mass of glycoprotein/glycan (presence of glycosylation)	gel filtration chromatography, mass profile by FAB/ES/MALDI-mass spectrometry, SDS/PAGE (before/after enzyme treatment)
Nature of carbohydrate-peptide linkage (N/O)	proteolytic digestion, amino acid analysis, examination of alkali lability, hydrazinolysis
Type of glycans (high-mannose, complex, hybrid), glycoforms	GLC-monosaccharide analysis, size/charge profile analysis, capillary electrophoresis
Number/proportions of glycans present	size/charge profile analyss, mapping by HPLC, HPAEC, FACE, MALDI-MS
Sequence of mono-saccharide residues	digestion by exoglycosidases, partial hydrolysis, NMR spectroscopy, mass spectrometry
Positions of glycosidic linkages	methylation analysis/GLC-MS, FAB-MS, NMR spectroscopy
Anomeric configuration	digestion by exoglycosidases, NMR spectroscopy
Certain structural determinants	antibody responses, endo/exoglycosidases, affinity chromatography/electrophoresis (lectins)
Type of charged substituents	size/charge profile analysis, HPLC, HPAEC, NMR spectroscopy
Spatial structure of glyco-protein/peptide/glycan	X-ray analysis, 2D/3D NMR spectroscopy, molecular dynamics, mechanics and modelling

about the type of structures that can be expected. The absolute configuration (D or L) of the monosaccharide residues can be established by GC of the corresponding (−)2-butyl glycosides [40].

Methylation analysis is a reliable chemical method for the elucidation of the substitution pattern and of the ring size of the individual monosaccharide residues in an oligosaccharide. After methylation [41] of all free hydroxyl groups, the monosaccharides liberated by hydrolysis are reduced, acetylated, and subsequently analyzed by GC-MS as partially O-methylated alditol acetates. The positions of the O-acetyl and O-methyl groups, as deduced from specific MS fragmentation, indicate the substitution pattern in the monosaccharide [42].

Information about the monosaccharide sequence in a glycan chain can be obtained enzymatically by successive exoglycosidase digestion [43]. In this approach monosaccharide residues are stepwise cleaved from the nonreducing terminus of the oligosaccharide by hydrolases (e.g. sialidase, α-mannosidase, β-galactosidase, etc.) which are highly specific towards their substrate including the anomeric configuration. By consequence, the change in effective size of the glycan can be analyzed by gel filtration chromatography under standard conditions before and after the enzyme treatment. The enzymatic sequence analysis of N-glycans has been automated into a RAAM (Reagent Array Analysis Method) GlycoSequencer [44].

Because in many instances only pico/nanomole amounts of material are available, much effort is focused on the development of profiling techniques based on high-performance separation procedures like capillary electrophoresis [45, 46], HPLC mapping [34, 47] and fluorophore assisted-carbohydrate electrophoresis (FACE) [48].

4.2 Mass spectrometry

Electron impact ionization mass spectrometry (EI-MS) in combination with GC-MS has already been used for a long time for the identification of monosaccharide derivatives in monosaccharide and methylation analysis [20]. This technique remains a standard for carbohydrate composition and linkage analysis. Chemical ionization mass spectrometry (CI-MS) can be used to increase sensitivity and detection of the molecular ion.

In recent years, several new mass spectrometric methods using soft ionization techniques like fast atom bombardment (FAB-MS), electrospray ionization (ES-MS) and matrix-assisted laser desorption ionization (MALDI-MS) have been applied for the study of glycoproteins, oligosaccharides and glycopeptides because of their high sensitivity. Structural information about branching pattern, number and length of branches and sequence in terms of hexoses, deoxyhexoses, N-acetylhexosamines and sialic acids, as well as the net molecular mass of the glycopeptide/protein, can be obtained.

FAB-MS is useful for defining the glycosylation sites at the polypeptide backbone and delineating the complete sequences of suitably derivatized individual glycans in glycopeptides from fragment ions [49, 50]. Glycopeptides and glycans have mostly been studied after permethylation. A typical example of FAB-mass ion fragmentation of an N- and O-glycan is given in Figure 4.

ES-MS has proven its power for the analysis of glycopeptides and glycoproteins, since the peptide/protein portion of the molecule usually provides the necessary multiple charge sites, ensuring mass-to-charge ratios within the range of the quadrupole analyzer. Interfacing ES mass spectrometers with liquid chromatography systems (e.g. HPLC-MS) allow rapid separation and analysis of mixtures of peptides and glycopeptides obtained after proteolytic digestion of a glycoprotein [51, 52]. Recent examples to highlight the applicability of MS technology are demonstrated by the structural analysis of the glycans of glycodelin-A, a human endometrial glycoprotein, glycodelin-S, a human seminal plasma glycoprotein [53], and horseradish

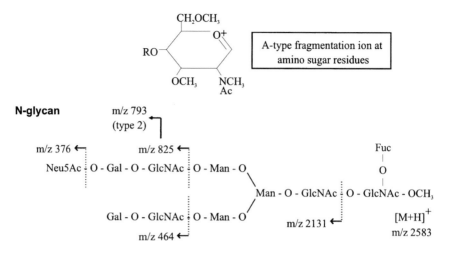

N-glycan

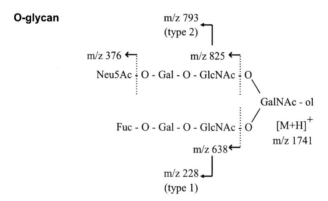

O-glycan

Figure 4. FAB-mass ion fragmentation of permethylated glycans derived from glycopeptides. A-type cleavage occurs reliably at each HexNAc residue, providing information about the composition of glycan moieties attached on the nonreducing sides of the cleavage sites.

peroxidase [54]. The introduction of array detectors and sector instruments increased the molecular ion detectability, thereby increasing the sensitivity to the picomole (femtomole) range.

The application of MALDI-TOF-MS (time-of-flight detection) is rapidly increasing in the glycoprotein/peptide field, due to its ease of operation [55] and its better tolerance of impurities (buffers, salts, additives and detergents) in comparison with other MS methods [56]. The recent introduction of delayed extraction technology for MALDI-MS has significantly improved mass accuracies [57, 58]. Producing only the molecular mass ion for each glycan, MALDI-MS is very useful for profiling glycan mixtures without derivatization. Furthermore, the new technique PSD (post-source decay)-MALDI-TOF-MS has been demonstrated to be a fast, highly sensitive and reproducible method for the localization of O-glyco-sylation sites of polymorphic epithelial mucin using glycopeptides [59] and the isomeric differentiation of N-glycans [60].

A development of particular significance in MS technology is the construction of a novel quadrupole orthogonal acceleration time-of-flight tandem mass spectrometer (Q-TOF) for ultra high sensitivity low femto-mole/attomole-range glycopeptide sequencing [61].

4.3 NMR spectroscopy

High-resolution NMR spectroscopy is the most powerful method for the unambiguous identification of N- as well as O-type carbohydrate chains. For the elucidation of the primary structure, it is the only (nondestructive) method that provides all details, comprising type of constituent mono-saccharides, including ring size and anomeric configuration, position of glycosidic linkages as well as position of noncarbohydrate substituents. Studying glycopeptides containing small peptide parts, usually the signals from the peptide portion do not interfere with those from the oligosaccha-ride. As an illustrative example, a typical ^{1}H-NMR spectrum of a glyco-peptide is shown in Figure 5.

Already at first sight, the one-dimensional (1D) ^{1}H-NMR spectrum can be used as a "fingerprint" containing many characteristic details to con-clude whether or not compounds are pure and/or identical. The definitive interpretation of the proton spectrum in terms of primary structure is based on the structural-report-group concept [62]. Most of the skeleton protons of the monosaccharide residues resonate in a narrow region between (δ) 3.5 and 3.9 ppm. Proton signals at clearly distinguishable positions outside this bulk, denoted as structural-reporter-group signals (Tab. 7), reveal

Table 7. ^{1}H-NMR structural-reporter-group signals in glycoprotein glycan analysis

- Anomeric protons (H-1)
- Amide protons (in H_2O)
- Protons attached to carbon atoms in the direct vicinity of a substitution position
    - Man H-2 and H-3
    - Gal H-3 and H-4
- Protons at deoxy carbon atoms
    - sialic acid H-3 (equatorial and axial)
    - Fuc H-5 and H-6 (CH_3)
- Protons shifted (out of the bulk region) due to glycosylation shifts
- Alditol protons
    - GalNAc-ol H-2, H-3, H-4 and H-5
- Protons shifted (out of the bulk region) due to noncarbohydrate substituents
    like alkyl, acyl, phosphate or sulfate groups
- Protons belonging to substituents on carbohydrate residues
    like O- and/or N-alkyl or acyl groups

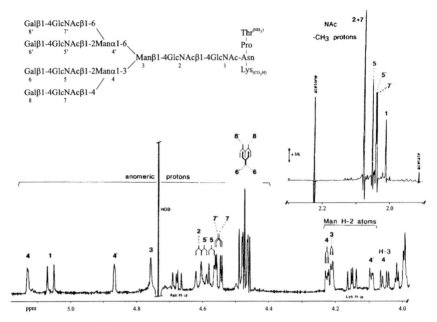

Figure 5. 500-MHz ¹H-NMR spectrum showing the structural-reporter-group regions of a desialylated complex-type glycan-peptide. The bulk region has been omitted. The boldface numbers in the spectrum refer to the corresponding residues in the structure. The relative-intensity scale of the NAc-proton region differs from that of the other part as indicated. The spectrum was recorded in D_2O at 300 K.

information on the primary structure depending on their chemical shifts (δ), intensities (area), line widths and coupling constants (J).

The application of various types of two-dimensional (2D)-NMR methods significantly improves the interpretation and assignments for the assessment of complex carbohydrate structures [63]. 2D ¹H-total correlation spectroscopy (TOCSY) is used to identify the spin systems in the constituent monosaccharides by means of the scalar coupled network starting from the anomeric protons. 2D ¹H-nuclear Overhauser effect spectroscopy (NOESY) is employed to obtain information on protons that are close in space, in particular the protons connected to the glycosidic linkage C atoms. It has to be noted that the strongest NOE is not always between these protons. Proton-carbon heteronuclear experiments correlate ¹H and ¹³C chemical shifts in a 2D spectrum. Heteronuclear multiple quantum-coherence spectroscopy (¹H-¹³C-HMQC) makes it possible to determine the correlation between carbon atoms and the directly attached protons. For determination of the position of the glycosidic linkage, heteronuclear multiple-bond correlation spectroscopy (HMBC) can be used. This method yields cross-peaks between ¹H and ¹³C atoms over several chemical bonds. Additionally, three-dimensional (3D)-NMR spectroscopy can be used for the investigation of the structure of complex carbohydrates. A nonselective 3D ^{1}H-^{13}C-NOE-TOCSY experiment and 3D HMQC-NOE spectroscopy provide detailed information on protons resonating in the bulk signal [64, 65].

The conversion of the NMR data into carbohydrate structures makes use of extensive libraries of reference compounds [66, 67] and computerized databases [68–70], now accessible through the internet [71].

Typical examples of glycan structure analyses as recently carried out in our research group include: the N-glycans of human urokinase [72], monoclonal immunoglobulin (Ig) G1 antibodies [73] and αD-hemocyanin [74], the N- and O-glycans of recombinant human erythropoietin [75], the O-glycans of equine chorionic gonadotropin [76], porcine zona pellucida glycoproteins [77], a bovine seminal plasma protein [78] and jacalin-bound rabbit IgG [79]. In all these cases the analyses, using ^{1}H-NMR spectroscopy as the major technique, were performed on the level of sialylated carbohydrate chains, providing also information about the sialylation patterns. The type of information obtainable from different NMR techniques is shown in Table 8.

Not only for the primary structural analysis but also for the study of the conformation and intramolecular interactions of glycans and glycoproteins/peptides, ^{1}H-NMR spectroscopy has proven to be an invaluable technique, providing data regarding the spatial structure and segmental mobility of carbohydrate chains in solution [80]. The solution conformation of oligosaccharides can be determined through the combined use of internuclear distances obtained from quantitative proton NOE measurements and tortional angles derived from coupling constants, in conjunction with molecular dynamics calculations. These data, together with those obtained from circular dichroism and X-ray diffraction, are crucial for computer-graphics molecular modelling of oligosaccharides and glycopeptides. Recently, some intact glycoproteins have been investigated by

Table 8. NMR parameters and their application for oligosaccharide structure determination

Relevant NMR parameter	Information
Chemical shift, spin couplings (^{1}H–^{1}H), spectral integration	monomer composition
Spin couplings (^{1}H–^{1}H, ^{13}C–^{1}H, ^{13}C–^{13}C)	monomer conformation
Chemical shift, spin couplings (^{1}H–^{1}H, ^{13}C–^{1}H)	anomeric configuration
Chemical shift, NOE, ROE, spin coupling (^{13}C–^{1}H)	linkage sites
Nuclear spin relaxation time (T_1, T_2), NOE, spin coupling (^{13}C–^{1}H, ^{13}C–^{13}C)	O-glycoside conformation
Spin couplings (^{1}H–^{1}H, ^{13}C–^{1}H)	hydroxymethyl conformation
Nuclear spin relaxation time (T_1, T_2), NOE	motional/dynamical properties

NMR [81, 82]. For instance, differences in conformational behavior of the unique N-glycan of pineapple stem bromelain glycoprotein with that of the bromelain-derived glycopeptide were demonstrated. For this purpose, models obtained through molecular dynamics simulations of the glyco-peptide in water were also applied [83, 84]. The two N-glycans of the free α-subunit of human chorionic gonadotropin (hCG) behave differently with reference to flexibility and interaction with the protein backbone as studied by ^{1}H-NMR spectroscopy [85].

5. Identification and analysis of glycosylation sites

Detailed knowledge of the carbohydrate structures at particular glycosyla-tion sites is not only a prerequisite for understanding their involvement in the biological action of certain glycoproteins, but it also provides informa-tion on the extent to which oligosaccharide biosynthesis may be affected by the surrounding protein structure within a given cellular glycosylation system [86]. Furthermore, new types of carbohydrate-protein linkages are still being discovered [87], for instance, the C-linkage of Man-Trp in human RNase2 [88, 89] and the occurrence of phosphoglycosylation [90].

In order to study site-specific glycosylation, a glycoprotein is either chemically and/or enzymatically cleaved into its peptides and glycopep-tides, which are then separated by a number of chromatographic steps. Recent examples are the separation of glycosylated peptides derived from recombinant tissue plasminogen activator [91] and recombinant coagula-tion factor VIIa [92] by use of sequential HPLC and capillary zone electro-phoresis.

A commonly applied strategy is the digestion of the glycoprotein with trypsin, chymotrypsin and/or V-8 protease followed by purification of glycopeptides representing individual N-glycosylation sites by RP-HPLC. After identification by amino acid sequence analysis, the glycopeptides are analyzed by ^{1}H-NMR spectroscopy and/or mass spectrometry. This procedure has successfully been applied to the determination of site-specific N-glycosylation of human chorionic gonadotropin [93], recom-binant human immunodeficiency virus (HIV)-1 gp120 [32], soybean seed coat peroxidase [94], recombinant bovine lactoferrin [95] and recombinant human interleukin-6 [96].

The analysis of the ratio of N-linked oligosaccharides attached to dif-ferent glycosylation sites of a glycoprotein requires the purification of site-specific glycopeptides. The proteolytic digestion of reduced and alkylated recombinant erythropoietin (r-EPO) to obtain site-specific glycopeptides was successfully performed with Lys-C and trypsin [97]. The determina-tion and characterization of the N-glycosylation sites of Tamm-Horsfall glycoprotein was performed on Glu-C(V-8 protease)-prepared glycopept-ides fractionated by RP-HPLC and Concanavalin A (Con A)-affinity chromatography. It was shown for the glycoprotein from one male donor that from the eight potential glycosylation sites seven sites were occupied and, remarkably, one of them (Asn251) contains predominantly high-mannose-type structures, $Man_{5-8}GlcNAc_2$ [98].

Recently developed mass spectrometric techniques have enabled the direct analysis of oligosaccharide chains while still attached to peptides, together with the peptide sequence. For instance, the reduced and alkylated glycoprotein is digested with a protease such as trypsin or endoproteinase Glu-C, and an aliquot of the digest is analyzed by HPLC connected direct-ly to an ES mass spectrometer to generate a "peptide mass map". Thus, molecular masses which do not match predicted proteolytically generated peptides are possibly glycopeptides. Tandem ES-MS/CID (collision-induc-ed dissociation)-MS can be used for selected ion monitoring (SIM) of carbohydrate-specific ions [99]. In order to help identify N-glycopeptides, the N-glycans are cleaved from the peptide by PNGase F or Endo H in an aliquot of the digest and rechromatographed to detect changes in the masses or retention times of the components in the digest. By combining the change in mass after different glycosidase treatments, the types of oligosaccharides on the glycopeptide can be determined [100]. Tandem mass spectrometry has also been proven to be useful for the analysis of O-glycopeptides, yielding the peptide sequence and the exact site of glyco-sylation on the peptide [101].

Recently, a new technique has been reported to gain site-specific com-positional data on the oligosaccharides attached to a single amino acid. Sequential solid-phase Edman degradation was used to recover a single glycosylated phenylthiohydantoin (PTH) amino acid from a (preferably desialylated) glycopeptide coupled to an arylamine membrane. The oligo-saccharide of the glyco-amino acid is characterized by HPAEC monosac-charide analysis and ES-MS [102, 103]. Although the yield is limited and the oligosaccharide is degraded slowly during repeated cycles of Edman degradation, the approach looks promising for the characterization of heavily O-glycosylated proteins.

6. Concluding remarks

The persistence throughout evolution of glycosylation of proteins, together with the fact that glycosylation of proteins is an expensive operation requir-ing genetic information for the production of many enzymes, substrates and cofactors, strongly indicates an important function for the variety of glycans. To understand the biological role of glycoprotein glycans, their detailed structure, conformation, and interactions with complementary molecules must be known. Furthermore, the complete characterization of glycoproteins will contribute to elucidating the biosynthetic controls that determine site-specific glycosylation patterns. In recent years the academic and industrial interests in the carbohydrate part of glycoproteins has grown dramatically, leading to new developments in separation techniques and in methodology to unravel the structure of complex carbohydrates. The progress in MS and NMR spectroscopy for the latter purpose is spectacular.

Acknowledgments
The studies in the authors' laboratory were supported by the Netherlands Foundation for Chemical Research (SON/NWO).

References

1 Montreuil J, Vliegenthart JFG, Schachter H (eds) (1995) *Glycoproteins: new comprehensive biochemistry*, vol. 29a. Elsevier Science, Amsterdam
2 Montreuil J, Vliegenthart JFG, Schachter H (eds) (1996) *Glycoproteins II: new comprehensive biochemistry*, vol. 29b. Elsevier Science, Amsterdam
3 Montreuil J, Vliegenthart JFG, Schachter H (eds) (1996) *Glycoproteins and disease: new comprehensive biochemistry*, vol. 30. Elsevier Science, Amsterdam
4 Allen HJ, Kisailus EC (eds) (1992) *Glycoconjugates: composition, structure and function.* Marcel Dekker, New York
5 Natsuka S, Lowe JB (1994) Enzymes involved in mammalian oligosaccharide biosynthesis. *Curr Opin Struct Biol* 4: 683–691
6 Vance BA, Wu W, Ribaudo RK, Segal DM, Kearse KP (1997) Multiple dimeric forms of human CD69 result from differential addition of N-glycans to typical (Asn-X-Ser/Thr) and atypical (Asn-X-Cys) glycosylation motifs. *J Biol Chem* 272: 23117–23122
7 Gavel Y, Von Heijne G (1990) Sequence differences between glycosylated and non-glycosylated Asn-X-Thr/Ser acceptor sites: implications for protein engineering. *Protein Engineering* 3: 433–442
8 Gooley AA, Classon BJ, Marschalek R, Williams KL (1991) Glycosylation sites identified by detection of glycosylated amino acids released from Edman degradation: the identification of Xaa-Pro-Xaa-Xaa as a motif for Thr-O-glycosylation. *Biochem Biophys Res Commun* 178: 1194–1201
9 Nehrke K, Ten Hagen KG, Hagen FK, Tabak LA (1997) Charge distribution of flanking amino acids inhibits O-glycosylation of several single-site acceptors *in vivo*. *Glycobiology* 7: 1053–1060
10 Lis H, Sharon N (1993) Protein glycosylation: structural and functional aspects. *Eur J Biochem* 218: 1–27
11 Bulet P, Hegy G, Lambert J, Van Dorsselaer A, Hoffmann JA, Hetru C (1995) Insect Immunity. The inducible antibacterial peptide Diptericin carries two O-glycans necessary for biological activity. *Biochemistry* 34: 7394–7400
12 Williams DH (1996) The glycopeptide story – how to kill the deadly "superbugs". *Nat Prod Rep* 13: 469–477
13 Vliegenthart JFG (1994) Studies on glycoprotein-derived carbohydrates. *Biochem Soc Trans* 22: 370–373
14 Vliegenthart JFG (1994) Studies on the carbohydrate chains of glycoproteins. In: Bock K, Clausen H (eds): *Complex carbohydrates in drug research.* Alfred Benzon Symposium 36: 30–41
15 Kamerling JP (1994) Structural studies on glycoprotein glycans. *Pure Appl Chem* 66: 2235–2238
16 Jenkins N, Parekh RB, James DC (1996) Getting the glycosylation right: implications for biotechnology industry. *Nature Biotechnol* 14: 975–981
17 Hounsel EF (ed) (1993) Glycoprotein analysis in biomedicine. *Methods in molecular biology*, vol. 14. Humana Press, Totowa, NJ
18 Sojar HT, Bahl OP (1987) Chemical deglycosylation of glycoproteins. *Methods Enzymol* 138: 341–350

19 Gerken TA, Gupta R, Jentoft N (1992) A novel approach for chemically deglycosylating O-linked glycoproteins. The deglycosylation of submaxillary and respiratory mucins. *Biochemistry* 31: 639–648

20 Kamerling JP, Vliegenthart JFG (1989) Mass Spectrometry. In: AM Lawson (ed): *Clinical biochemistry: principles, methods, applications*, vol. 1. Walter de Gruyter, Berlin, 176–263

21 Townsend RR, Hardy MR (1991) Analysis of glycoprotein oligosaccharides using high-pH anion exchange chromatography. *Glycobiology* 1: 139–147

22 Dwek RA, Edge CJ, Harvey DJ, Dormald MR, Parekh RB (1993) Analysis of glycoprotein-associated oligosaccharides. *Annu Rev Biochem* 62: 65–100

23 Takasaki S, Mizuochi T, Kobata A (1982) Hydrazinolysis of asparagine-linked sugar chains to produce free oligosaccharides. *Methods Enzymol* 83: 263–268

24 Merry AH, Bruce J, Bigge C, Ioannides A (1992) Automated simultaneous release of intact and unreduced N- and O-linked glycans from glycoproteins. *Biochem Soc Trans* 20: 91s

25 Schneider P, Ferguson MAJ (1995) Microscale analysis of glycosylphosphatidylinositol structures. *Methods Enyzmol* 250: 614–630

26 Nilsson B, Zopf D (1982) Gas chromatography and mass spectrometry of hexosamine-containing oligosaccharide alditols as their permethylated *N*-trifluoroacetyl derivatives. *Methods Enzymol* 83: 46–58

27 Zinn AB, Plantner JJ, Carlson DM (1977) Nature of linkages between protein and oligosaccharides. In: MI Horowitz, W Pigman (eds): *The glycoconjugates*. Academic Press, New York, 69–85

28 Tarentino AL, Plummer TH (1994) Enzymatic deglycosylation of asparagine-linked glycans: purification, properties, and specificity of oligosaccharide-cleaving enzymes from *Flavobacterium meningosepticum. Methods Enzymol* 230: 44–57

29 Ishii-Karakasa I, Iwase H, Hotta K, Tanaka Y, Omura S (1992) Partial purification and characterization of an endo-α-*N*-acetylgalactosaminidase from the culture medium of *Streptomyces* sp. OH-11242. *Biochem J* 288: 475–482

30 Damm JBL, Kamerling JP, Van Dedem GWK, Vliegenthart JFG (1987) A general strategy for the isolation of carbohydrate chains from N,O-glycoproteins and its application to human chorionic gonadotropin. *Glycoconjugate J* 4: 129–144

31 Finne J, Krusius T (1982) Preparation and fractionation of glycopeptides. *Methods Enzymol* 83: 269–277

32 Yeh J, Seals JR, Murphy CI, Van Halbeek H, Cummings RD (1993) Site-specific N-glycosylation and oligosaccharide structures of recombinant HIV-1 gp120 derived from a *Baculovirus* expression system. *Biochemistry* 32: 11087–11099

33 Hunter AP, Games DE (1995) Evaluation of glycosylation site heterogeneity and selective identification of glycopeptides in proteolytic digests of bovine α_1-acid glycoprotein by mass spectrometry. *Rapid Commun Mass Spectrom* 9: 42–56

34 Rohrer JS, Cooper GA, Townsend RR (1993) Identification, quantification and characterization of glycopeptides in reversed-phase HPLC separations of glycoprotein proteolytic digests. *Anal Biochem* 212: 7–16

35 Davies MJ, Smith KD, Carruthers RA, Chai W, Lawson AM, Hounsell EF (1993) Use of a porous graphitised carbon column for the high-performance liquid chromatography of oligosaccharides, alditols and glycopeptides with subsequent mass spectrometry analysis. *J Chromatog* 646: 317–326

36 Hsi KL, Chen L, Hawke DH, Zieske LR, Yuan PM (1991) A general approach for characterizing glycosylation sites of glycoproteins. *Anal Biochem* 198: 238–245

37 Osawa T, Tsuji T (1987) Fractionation and structural assessment of oligosaccharides and glycopeptides by use of immobilized lectins. *Annu Rev Biochem* 56: 21–42

38 Merkle RK, Cummings RD (1987) Lectin affinity chromatography of glycopeptides. *Methods Enzymol* 138: 232–259

39 Shimura K, Kasai K (1997) Affinity capillary electrophoresis: A sensitive tool for the study of molecular interactions and its use in microscale analysis. *Anal Biochem* 251: 1–16

40 Gerwig GJ, Kamerling JP, Vliegenthart JFG (1979) Determination of the absolute configuration of monosaccharides in complex carbohydrates by capillary GLC. *Carbohydr Res* 77: 1–7

41 Jay A (1996) The methylation reaction in carbohydrate analysis. *J Carbohydr Chem* 15: 897–923

42 Hellerqvist CG, Sweetman BJ (1990) Mass spectrometry of carbohydrates. Methods Biochem Anal 34: 91–143

43 Yamashita K, Mizuochi T, Kobata A (1982) Analysis of oligosaccharides by gel filtration. *Methods Enzymol* 83: 105–126

44 Edge CJ, Rademacher TW, Wormald MR, Parekh RB, Butters TD, Wing DR, Dwek RA (1992) Fast sequencing of oligosaccharides: the Reagent Array Analysis Method. *Proc Natl Acad Sci USA* 89: 6338–6342

45 Hermentin P, Doenges R, Witzel R, Hokke CH, Vliegenthart JFG, Kamerling JP, Conradt HS, Mimtz M, Brazel D (1994) A strategy for the mapping of N-glycans by high-performance capillary electrophoresis. *Anal Biochem* 221: 29–41

46 Guttmann A, Starr CM (1995) Capillary and slab gel electrophoresis of oligosaccharides. *Electrophoresis* 16: 993–997

47 Takahashi N, Nakagawa H, Fujikawa K, Kawamura Y, Tomiya N (1995) Three-dimensional elution mapping of pyridylaminated N-linked neutral and sialyl oligosaccharides. *Anal Biochem* 226: 139–146

48 Starr C, Masada RI, Hague C, Skop E, Klock J (1996) Fluorophore-Assisted-Carbohydrate-Electrophoresis, FACE, in the separation, analysis, and sequencing of carbohydrates. *J Chromatography* 720: 295–321

49 Dell A (1989) FAB mass spectrometry of carbohydrates. *Adv Carbohydr Chem Biochem* 45: 19–72

50 Khoo KH, Chatterjee D, Caulfield JP, Morris HR, Dell A (1997) Structural mapping of the glycans from egg glycoproteins of *Schistosoma mansoni* and *Schistosoma japonicum*. identification of novel core structures and terminal sequences. *Glycobiology* 7: 663–677

51 Medzihradszky KF, Maltby DA, Hall SC, Settineri CA, Burlingame AL (1994) Characterization of protein N-glycosylation by reversed-phase microbore liquid chromatography/electrospray mass spectrometry, complementary mobile phases, and sequential exoglycosidase digestion. *J Am Soc Mass Spectrom* 5: 350–358

52 Settineri CA, Burlingame AL (1995) Mass spectrometry of carbohydrates and glycoconjugates. In: El Rassi Z (ed): *Carbohydrate analysis: high performance liquid chromatography and capillary electrophoresis*. Elsevier Science, Amsterdam, 447–514

53 Morris HR, Dell A, Easton RL, Panico M, Koiestinen H, Koistinen R, Oehninger S, Patankar MS, Seppala M, Clark GF (1996) Gender-specific glycosylation of human glycodelin affects its contraceptive activity. *J Biol Chem* 271: 32 159–32 170

54 Takahashi N, Lee KB, Nakagawa H, Tsukamoto Y, Masuda K, Lee YC (1998) New N-glycans in horseradish peroxidase. *Anal Biochem* 255: 183–187

55 Treuheit MJ, Costello CE, Halsall HB (1992) Analysis of the five glycosylation sites of human α_1-acid glycoprotein. *Biochem J* 283: 105–112

56 Vorm O, Roepstorff P, Mann M (1994) Improved resolution and very high sensitivity in MALDI-TOF of matrix surfaces made by fast evaporation. *Anal Chem* 66: 3281–3287

57 Vestal ML, Juhasz P, Martin SA (1995) Delayed extraction matrix-assisted laser desorption time-of-flight mass spectrometry. *Rapid Commun Mass Spectrom* 9: 1044–1050

58 Sugiyama E, Hara A, Uemura K, Taketomi T (1997) Application of matrix-assisted laser desorption ionization time-of-flight mass spectrometry with delayed ion extraction to ganglioside analysis. *Glycobiology* 7: 719–724

59 Goletz S, Thiede B, Hanisch F-G, Schultz M, Peter-Katalinic J, Muller S, Seitz O, Karsten U (1997) A sequencing strategy for the localization of O-glycosylation sites of MUC1 tandem repeats by PSD-MALDI mass spectrometry. *Glycobiology* 7: 881–896

60 Rouse JC, Strang A-M, Yu W, Vath JE (1998) Isomeric differentiation of asparagine-linked oligosaccharides by matrix-assisted laser desorption-ionization postsource decay time-of-flight mass spectrometry. *Anal Biochem* 256: 33–46

61 Morris HR, Paxton T, Panico M, McDowell R, Dell A (1997) A novel geometry mass spectrometer, the Q-TOF, for low-femtomole/attomole-range biopolymer sequencing. *J Protein Chem* 16: 469–479

62 Vliegenthart JFG, Van Halbeek H, Dorland L (1981) The application of 500-MHz high-resolution ^{1}H-NMR spectroscopy for the structural determination of carbohydrates derived from glycoproteins. *Pure Appl Chem* 53: 45–77

63 Hård K, Vliegenthart JFG (1993) Nuclear magnetic resonance spectroscopy of glyco-protein-derived carbohydrate chains. In: M Fukuda, A Kobata (eds): *Glycobiology, a practical approach*. Oxford University Press, Oxford, 223–242

64 Vuister GW, De Waard P, Boelens R, Vliegenthart JFG, Kaptein R (1989) The use of 3D NMR in structural studies of oligosaccharides. *J Am Chem Soc* 111: 772–774

65 De Waard P, Vuister R, Boelens R, Vliegenthart JFG (1990) Structural studies by ^{1}H/^{13}C two-dimensional and three-dimensional HMQC-NOE of natural abundance in complex carbohydrates. *J Am Chem Soc* 112: 3232–3234

66 Vliegenthart JFG, Dorland L, Van Halbeek H (1983) High-resolution, ^{1}H-nuclear mag-netic resonance spectroscopy as a tool in the structural analysis of carbohydrates related to glycoproteins. *Adv Carbohydr Chem Biochem* 41: 209–374

67 Kamerling JP, Vliegenthart JFG (1992) High-resolution ^{1}H-nuclear magnetic resonance spectroscopy of oligosaccharide-alditols released from Mucin-type O-glycoproteins. *Biol Magn Reson* 10: 1–194

68 Van Kuik JA, Hård K, Vliegenthart JFG (1992) A ^{1}H-NMR database computer program for the analysis of the primary structure of complex carbohydrates. *Carbohydr Res* 235: 53–68

69 Van Kuik JA, Vliegenthart JFG (1992) Databases of complex carbohydraes. *Trends Bio-technol* 10: 182–185

70 Van Kuik JA, Vliegenthart JFG (1994) A NMR spectroscopic database of complex carbo-hydrate structures. *Carbohydr Europe* 10: 31–32

71 URL: http://www.boc.chem.uu.nl/sugabase/databases.html

72 Bergwerff AA, Van Oostrum J, Kamerling JP, Vliegenthart JFG (1995) The major N-linked carbohydrate chains from human urokinase; the occurrence of 4-O-sulfate, α(2-6)-sialylated or α(1-3)-fucosylated *N*-acetylgalactosamine-β(1-4)-*N*-acetylglucosamine ele-ments. *Eur J Biochem* 228: 1009–1019

73 Bergwerff AA, Stroop CJM, Murray B, Holtorf AP, Pluschke G, Van Oostrum J, Kamer-ling JP, Vliegenthart JFG (1995) Variation in N-linked carbohydrate chains in different batches of two chimeric monoclonal IgG1 antibodies produced by different murine SP2/0 transfectoma cell subclones. *Glycoconjugate J* 12: 318–330

74 Lommerse JPM, Thomas-Oates JE, Gielens C, Preaux G, Kamerling JP, Vliegenthart JFG (1997) Primary structure of 21 novel monoantennary and diantennary N-linked carbo-hydrate chains from α D-hemocyanin of *Helix pomatia*. *Eur J Biochem* 249: 195–222

75 Hokke CH, Bergwerff AA, Van Dedem GWK, Kamerling JP, Vliegenthart JFG (1995) Structural analysis of sialylated N- and O-linked carbohydrate chains of recombinant human erythropoietin expressed in Chinese hamster ovary cells; sialylation patterns and branch location of dimeric *N*-acetyllactosamine units. *Eur J Biochem* 228: 981–1008

76 Hokke CH, Roosenboom MJH, Thomas-Oates JE, Kamerling JP, Vliegenthart JFG (1994) Structure determination of the disialylated poly-(*N*-acetyllactosamine)-containing O-link-ed carbohydrate chains of equine chorionic gonadotropin. *Glycoconjugate J* 11: 35–41

77 Hokke CH, Damm JBL, Penninkhof B, Aitken RJ, Kamerling JP, Vliegenthart JFG (1994) Structure of the O-linked carbohydrate chains of procine zona pellucida glycoproteins. *Eur J Biochem* 221: 491–512

78 Gerwig GJ, Calvete JJ, Töpfer-Petersen E, Vliegenthart JFG (1996) The structure of O-linked carbohydrate chains of bovine seminal plasma protein PDC-109 revised by ^{1}H-NMR spectroscopy. A correction. *FEBS Lett* 387: 99–100

79 Kabir S, Gerwig GJ (1997) The structural analysis of the O-glycans of the jacalin-bound rabbit immunoglobulin G. *Biochem Molec Biol Int* 42: 769–778

80 Van Halbeek H (1994) NMR developments in structural studies of carbohydrates and their complexes. *Curr Opin Struct Biol* 4: 697–709

81 Fletcher MC, Harrison RA, Lachmann PJ, Neuhaus D (1994) Structure of a soluble, glyco-sylated form of the human complement regulatory protein CD59. *Structure* 2: 185–199

82 De Beer T, Van Zuylen CWEM, Leeflang BR, Hård K, Boelens R, Kaptein R, Kamerling JP, Vliegenthart JFG (1996) NMR studies of the free α subunit of human chorionic gona-dotropin. Structural influences of N-glycosylation and the β subunit on the conformation of the α subunit. *Eur J Biochem* 241: 229–242

83 Lommerse JPM, Kroon-Batenburg LMJ, Kamerling JP, Vliegenthart JFG (1995) Con-formational analysis of the xylose-containing N-glycan of pineapple stem bromelain as part of the intact glycoprotein. *Biochemistry* 34: 8196–8206

84 Lommerse JPM, Kroon-Batenburg LMJ, Kroon J, Kamerling JP, Vliegenthart JFG (1995)
    Conformations and internal mobility of a glycopeptide derived from bromelain using
    molecular dynamics simulations and NOESY analysis. *J Biomol NMR* 5:79–94
85 Van Zuylen CWEM, Kamerling JP, Vliegenthart JFG (1997) Glycosylation beyond the
    Asn78-linked GlcNAc residues has a significant enhancing effect on the stability of the α
    subunit of human chorionic gonadotropin. *Biochem Biophys Res Commun* 232: 117–120
86 Yet MG, Wold F (1990) The distribution of glycan structures in individual N-glycosyla-
    tion sites in animal and plants glycoproteins. *Arch Biochem Biophys* 278: 356–364
87 Hayes BK, Hart GW (1994) Novel forms of protein glycosylation. *Curr Opin Struct Biol*
    4: 692–696
88 Hofsteenge J, Müller DR, De Beer T, Löffler A, Richter WJ, Vliegenthart JFG (1994)
    New type of linkage between a carbohydrate and a protein: C-glycosylation of a specific
    tryptophan residue in human RNase Us. *Biochemistry* 33: 13 524–13 530
89 Hofsteenge J, Löffler A, Müller DR, Richter WJ, De Beer T, Vliegenthart JFG (1996)
    Protein C-glycosylation. *Techniques Protein Chem* VII: 163–171
90 Haynes PA (1998) Phosphoglycosylation: a new structural class of glycosylation? *Glyco-
    biology* 8: 1–5
91 Wu SL (1997) The use of sequential high-performance liquid chromatography and capil-
    lary zone electrophoresis to separate the glycosylated peptides from recombinant tissue
    plasminogen activator to a detailed level of microheterogeneity. *Anal Biochem* 253: 85–97
92 Weber PL, Kornfelt T, Klausen NK, Lunte SM (1995) Characterization of glycopeptides
    from recombinant coagulation factor VIIa by high-performance liquid chromatography
    and capillary zone electrophoresis using ultraviolet and pulsed electrochemical detection.
    Anal Biochem 225: 135–142
93 Weisshaar G, Hiyama J, Renwick AGC (1991) Site-specific N-glycosylation of human
    chorionic gonadotropin-structural analysis of glycopeptides by one- and two-dimensional
    ¹H-NMR spectroscopy. *Glycobiology* 1: 393–404
94 Gray JSS, Montgomery R (1997) The N-glycosylation sites of soybean seed coat perox-
    idase. *Glycobiology* 7: 679–685
95 Lopez M, Coddeville B, Langridge J, Plancke Y, Sautiere P, Chaabihi H, Chirat F, Harduin-
    Lepers A, Cerutti M, Verbert A, Delannoy P (1997) Microheterogeneity of the oligo-
    saccharides carried by recombinant bovine lactoferrin expressed in *Mamestra brassicae*
    cells. *Glycobiology* 7: 635–651
96 Orita T, Oh-eda M, Hasegawa M, Kuboniwa H, Esaki K, Ochi N (1994) Polypeptide and
    carbohydrate structure of recombinant human interleukin-6 produced in Chinese hamster
    ovary cells. *J Biochem* 115: 345–350
97 Sasaki H, Ochi N, Dell A, Fukuda M (1988) Site-specific glycosylation of human recom-
    binant erythropoietin: analysis of glycopeptides or peptides at each glycosylation site by
    fast atom bombardment mass spectrometry. *Biochemistry* 27: 8618
98 Van Rooijen JJM, Voskamp AF, Kamerling JP, Vliegenthart JFG (1999) Glycosylation
    sites and site-specific glycosylation in human Tamm-Horsfall glycoprotein. *Glycobiology*
    9: 21–30
99 Reinhold VN, Reinhold BB, Costello CE (1995) Carbohydrate molecular weight profil-
    ing, sequence, linkage, and branching data: ES-MS and CID. *Anal Chem* 67: 1772–1784
100 Carr SA, Hemling ME, Bean MF, Roberts GD (1991) Integration of mass spectrometry in
    analytical biotechnology. *Anal Chem* 63: 2802–2824
101 Medzihradszky KF, Gillece-Castro BL, Settineri CA, Townsend RR, Masiarz FR, Bur-
    lingame AL (1990) Structure determination of O-linked glycopeptides by tandem mass
    spectrometry. *Biomed Environ Mass Spectrom* 19: 777–781
102 Gooley AA, Pisano A, Packer NH, Ball M, Jones A, Alewood PF, Redmond JW, Williams
    KL (1994) Characterization of a single glycosylated asparagine site on a glycopeptide
    using solid-phase Edman degradation. *Glycoconjugate J* 11: 180–186
103 Gooley AA, Williams KL (1997) How to find, identify and quantitate the sugars on pro-
    teins. *Nature* 385: 557–559

Proteomics in Functional Genomics
ed. by P. Jollès and H. Jörnvall
© 2000 Birkhäuser Verlag Basel/Switzerland

Lipopeptide preparation and analysis

Jan Johansson*, Margareta Stark, Magnus Gustafsson,
Yuqin Wang and Shahparak Zaltash

*Department of Medical Biochemistry and Biophysics, Karolinska Institutet,
SE-171 77 Stockholm, Sweden*

Summary. Lipophilic peptides and proteins present specific problems during preparation and analysis which require the use of modified methodology. This chapter discusses some of the methods that have been employed in the isolation and structural studies of the pulmonary surfactant-associated proteins B and C (SP-B and SP-C), other proteins with lipidlike physico-chemical properties, and the SP-B precursor. In particular, methods for separation and analysis of peptide/lipid mixtures, high-resolution separation of lipopeptides, analysis of fatty acylated peptides, and secondary and tertiary structure analysis of lipopeptides are discussed.

Introduction

Biochemical studies of nonpolar polypeptides, lipopeptides, was introduced by Folch and Lees who described membrane proteins which are soluble in chloroform/methanol mixtures and insoluble in water [1]. These proteins are hydrophobic even after removal of associated lipids, and lipopeptides are as a rule membrane-associated. Lipopeptides occur as membrane components in many animal, plant and bacterial cells and exhibit physico-chemical properties similar to those of lipids, and are thus not soluble in the aqueous buffers normally used for protein analysis. Protein association to membranes can be brought about via several mechanisms, including nonpolar transmembrane segments, the presence of covalently attached lipid moieties interacting with the membrane interior, and ionic interactions between protein basic residues and negatively charged phospholipids. The nonpolar segments and the covalently bound lipids pose special problems in the analysis of lipopeptides.

Pulmonary surfactant is a complex mixture of mainly phospholipids and small amounts of proteins (about 10% of the surfactant mass). The main function of this system is to reduce the alveolar surface tension by forming a monolayer of phospholipids at the air/liquid interface. The surfactant proteins B and C (SP-B and SP-C) are lipopeptides which are thought to be important for spreading of the phospholipids to the interface [2]. SP-B and SP-C were purified to apparent homogeneity from chloroform/methanol

---

* Corresponding author.

extracts of lung tissue by liquid-gel chromatography over Lipidex-5000 in
ethylene chloride/methanol, followed by gel filtration over Sephadex
LH-60 in chloroform/methanol containing 0.1 M HCl [3]. This purification scheme with recent modifications has subsequently been used for
isolation of significant amounts of SP-B and SP-C for structural and
functional studies [2] as well as for isolation of lipopeptides from other
tissues.

1. Isolation of lipopeptides

For isolation of lipopeptides from lipid-rich sources like lung, spinal cord
and bile, organic extraction followed by reverse-phase (RP) chromatography
on Lipidex-5000 in organic solvents can be used as the initial purification step [3, 4]. Tissue material is homogenized in chloroform/methanol,
2:1 (v/v), after which 0.25 vol 0.1 M NaCl in water is added, forming the
two-phase system chloroform/methanol/0.1 M NaCl, 8:4:3 (by vol). After
phase separation, the lower (organic) phase, containing lipids and lipid-
associated proteins, is evaporated to dryness, redissolved and subjected to
separation by RP chromatography on Lipidex-5000 in the solvent system
methanol/ethylene chloride, 4:1 (v/v) [3]. In this system the phospholipids
elute in the first Lipidex-5000 column volume, while neutral lipids elute
significantly later [5]. In the original scheme for purification of SP-B and
SP-C from each other and from the phospholipids, the contents of the entire
first column volume were collected and subjected to Sephadex LH-60 chromatography in chloroform/methanol, 1:1 (v/v) containing 5% 0.1 M HCl
[6]. This yields apparently homogeneous SP-B and SP-C preparations [3].
    The strategy to collect the entire first column volume suggests that there
is no significant separation of the lung surfactant lipopeptides and phospholipids by Lipidex-5000 chromatography, an assumption that previously
was difficult to test because of problems associated with analysis of protein
contents in the presence of large amounts of lipids. We have recently overcome these problems, and the protein content in Lipidex-5000 fractions can
now be determined by amino acid analysis. For this purpose the hydrolyzed
samples are treated with phenylisothiocyanate, and the phenylthiocarbamyl
(PTC) derivatives obtained are separated on a C18 RP column using a
phosphate buffer/acetonitrile gradient, and detected by their absorbance at
254 nm [7]. With this method samples containing up to 95% (w/w) phospholipids can be analyzed without affecting the yields of any of the amino
acids [8]. PITC reacts also with the primary amino group present in phosphatidylethanolamine and phosphatidylserine, which can therefore also be
determined, along with the proteins. With the PTC/RP-HPLC method, lipid
contamination of the analytical HPLC column is minimized by washing
with 60% aqueous acetonitrile between each run, and with methanol/
isopropanol, 1:1 (v/v), after each 30–50 runs. This largely reduces pre-

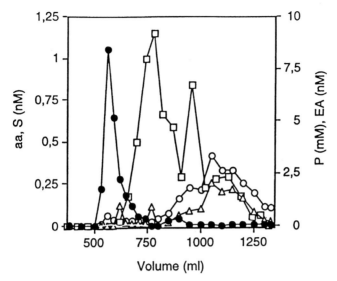

Figure 1. Separation of lipopeptides and phospholipids. Protein and phospholipid profiles after Lipidex-5000 chromatography of an organic lung extract. Protein (determined as total amino acids, aa, ●); serine (Ser, ○); ethanolamine (EA, △) were determined by the PTC/RP-HPLC method. 200-µl aliquots of 10-ml fractions were hydrolyzed and derivatized and 5% analyzed by RP-HPLC. Total phospholipids were determined by phosphorous analysis (P, □). Column volume was 1320 ml.

vious problems associated with consecutive analysis of extremely lipid-rich samples using the classical ninhydrin-based amino acid analysis method. Practical problems then related to difficulties in removing lipids from the ion-exchange column, which could not be regenerated easily. The advantage with the PTC/RP-HPLC method is thus that (i) amino acids and amino-group-containing phospholipids can be determined at the same time, thereby reducing the need for a separate phosphorous analysis during purification of proteins from lipid-rich sources, and that (ii) hundreds of samples can be analyzed using the same column [8].

In all tissues that we have examined (bile, lung, spinal cord, neutrophil granulocytes) for protein and phospholipid profiles after Lipidex-5000 chromatography, the polypeptides elute significantly earlier than the phospholipids (Fig. 1). Thus, Lipidex-5000 chromatography in methanol/ ethylene chloride separates proteins and phospholipids present in organic solvent extracts of tissue material. This separation may be advantageous for the isolation of novel lipopeptides in organic tissue extracts, since the protein fractions can be directly resolved by RP-HPLC (see below), thereby reducing the need for separate removal of phospholipids, by e.g. Sephadex LH-60 chromatography.

2. RP-HPLC of lipopeptides

RP-HPLC is an attractive method for high-resolution separation of lipopeptides, but the pronounced hydrophobicity of lipopeptides requires the
use of modified solvent systems. Recently, we developed an RP-HPLC
procedure for analysis of SP-B and SP-C which uses a C18 resin as stationary phase, aqueous methanol or ethanol as initial solvents, and a linear
gradient of 2-propanol for elution [9]. This system has several features that
makes it attractive for isolation and analysis of lipopeptides. Thus:

- Since the initial solvent mixture contains low amounts of water (down to
  5%) and the samples are injected in water-free alcohol, aggregation of
  lipopeptides is kept at a minimum.
- High resolution is achieved; e.g. SP-C peptides containing two and
  three palmitoyl groups can be separated. If the chromatography is performed at elevated temperature, impressive resolution is obtained; at
  70°C C-terminally methylated and nonmethylated forms of the 4.2-kDa
  SP-C separate.
- Conformation-dependent separation is possible, as judged from the different elution patterns of SP-C in α-helical and β-sheet conformation
  [9]. A nonhelical SP-C analogue elutes as a much broader peak than the
  native peptide, which yields a sharp symmetric peak. A broad peak is
  in agreement with a heterogeneous mixture of aggregated peptides. Circular dichroism (CD) spectroscopy of peptides eluting under the broad
  peak confirms that they have β-sheet structures. HPLC thus affords a
  mode to analyze the gross secondary structure of lipopeptides. Furthermore, separation of two 21-residue peptides with identical amino acid
  compositions but with different distributions of polar and nonpolar residues is observed (Fig. 2). Provided that the peptides exhibit helical
  conformations, the early eluting peptide has a mixed nonpolar/polar
  surface, whereas the late eluting peptide is amphiphilic, i.e. the nonpolar
  and polar residues are clustered in different regions of the helical circum-
  ference. We have studied the peptides using CD and Fourier transform
  infrared (FTIR) spectroscopy, which clearly shows that they are α-heli-
  cal in organic solvents and phospholipid bilayers and micelles [10, 11].
  The results from HPLC analysis of these peptides (Fig. 2) thus indicate
  that they are helical under the RP chromatography conditions, and that
  the concomitant different surfaces of the peptides interact differently
  with the stationary phase. Qualitatively similar results have been obtained with other, more polar, model peptides [12].

Using this HPLC system, a novel isoform of SP-C with three covalently
linked palmitoyl chains was identified [9]. This extremely nonpolar peptide
contains two thioester-linked pamitoyl chains and one palmitoyl chain
amide-linked to the side-chain amino group of a lysine residue (Fig. 3).
Edman degradation of the tripalmitoylated isoform reveals that the likely

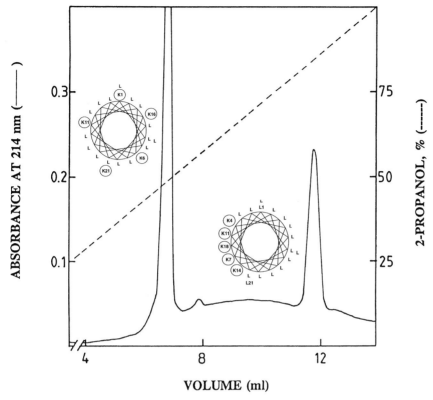

Figure 2. Reversed-phase HPLC of lipopeptides. Chromatogram showing the separation of two peptides with identical amino acid composition (K_5L_{16}) but with different helical surfaces. Next to each elution peak a helical wheel diagram is shown, in which the Lys residues of the corresponding peptide are circled.

location of the acyl chain is at Lys11. This was concluded from the facts that the α-amino group is free, at cycle 11 only trace amount of phenylthiohydantion-Lys was detected, and no other residues with reactive side chains are present in SP-C. The existence of a tripalmitoylated SP-C isoform has recently been confirmed by electrospray ionization (ESI) mass spectrometry [13]. It was suggested that the N^ε-palmitoyl group may be non-enzymatically formed by acyl transfer from neighbouring phospholipids to membrane-bound SP-C [9].

We believe that the possibility to separate very nonpolar lipopeptides in combination with the high resolution obtained with the HPLC system may enable isolation of lipopeptides present in organic tissue extracts which previously have escaped detection because of low abundance and/or extreme hydrophobicity.

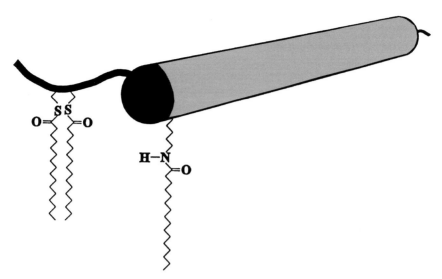

Figure 3. Tripalmitoylated SP-C. Schematic picture of a tripalmitoylated isoform of SP-C where the two thioester-linked and single amide-bound palmitoyl moieities are highlighted. The barrel represents the SP-C α-helix covering residues 9–34.

3. Analysis of acylated lipopeptides

Several proteins carry covalently linked lipid moieties. There are at least four different modes of covalent attachment of lipids to proteins: myristoylation of the α-amino group, fatty acylation of Cys thiol groups via thioester-linked palmitoyl or other long-chain fatty acids (mainly stearic acid), prenylation of C-terminally located Cys residues via thioether-linked farnesyl, geranyl-geranyl, or dolichyl groups, and linkage of glycosylphosphatidylinositol anchors to the C-terminal carboxy group [14, 15]. The sequence specificity and enzymology of lipid attachment is comparatively well known in the cases of myristoylation, prenylation, and glycosylphosphatidylinositol anchoring, and many of the proteins that carry these groups are dependent on the lipid moieties for membrane interaction, i.e. they are not lipopeptides in the true sense. In contrast, in the case of thioester-linked fatty acylation, the enzymes involved, if any (see below), as well as the structural and functional consequences, are largely unknown, and several acylated proteins remain membrane-associated also after removal of the acyl chains. These circumstances make studies of structure/activity relationships of acylated lipopeptides important.

Often, the presence of thioester-linked acyl chains is deduced from incorporation of radioactive acyl chains, as detected by gel electrophoresis and autoradiography, in combination with analysis of the stability of the chemical bond responsible for the linkage. The latter can distinguish between amide-bound myristoyl groups and thioester-bound palmitoyl

Figure 4. Mass spectrometric sequence analysis of SP-C. Survey of SP-C fragments obtained with electrospray ionization/collision-induced dissociation. The upper line shows the SP-C amino acid sequence in one-letter code, where Cys5 and 6 are palmitoylated. Below that, the fragments are identified with their respective start and stop locations. Solid lines represent fragments that could be unequivocally assigned to specific peptide segments. Dotted lines represent fragments where several assignments are possible, in which case each line corresponds to one of the segments possible.

groups [16], but the exact sequence location is difficult to establish with these techniques. Determination of protein acylation sites by direct analysis has proved to be fruitful in several cases [14, 17–19]. The intrinsic lability of the thioester bond requires special precautions during analysis of peptides carrying fatty acyl chains [17]. SP-C is palmitoylated at Cys5 and 6 in all species analyzed except for dog and mink, which are monopalmitoylated and have one Cys only [18, 19]. This was deduced from a combination of Edman sequence analysis, time-of-flight mass spectrometry, and analysis of the susceptibility of the bound acyl chains to treatment with basic and reducing agents [18]. Recently, we have analyzed SP-C using ESI mass spectrometry (Fig. 4), which has confirmed the amino acid sequence, including the presence of the covalently linked palmitoyl groups [13]. ESI mass spectrometry appears to be a very useful method for sequence analysis of lipopeptides since the presence of covalent modifications of the polypeptide chain is obvious (provided that the amino acid sequence is known), mixtures of organic solvents are suitable for spraying, and the amount of peptide consumed is small compared with that with most other techniques.

The covalently linked palmitoyl groups of SP-C may be important for the structure of the peptide; deacylation yields a decrease in α-helical content with about 20% [20–22]. The acyl chains have also been reported to be crucial for surface activity and lipid film stability of SP-C/lipid mixtures [22, 23]. SP-C isolated from the lung lavage of patients with alveolar proteinosis is modified by partial or complete removal of the palmitoyl residues, possibly contributing to a reduced surfactant function [24]. The mechanisms whereby the palmitoyl groups modulate SP-C structure and function are not known, and different possibilities have been suggested. These include N-capping of the first turn of the α-helix [25] and effects on the orientation of SP-C in a phospholipid monolayer [26]. In order to study the effects of the palmitoyl groups, we have synthesized palmitoylated model peptides using a recently developed method for *S*- and *O*-palmitoylation [27]. Typically, 2–10 mg of peptide is dissolved in 100–500 µl anhydrous acid, and acyl chloride is added. The reaction is quenched by addition of aqueous ethanol. Acylated peptides can be purified by chromatography over Lipidex-5000 in ethylene chloride/methanol, 1:4 (v/v), and with the RP-HPLC system optimized for lipopeptide separation (see above). Lipopeptides and palmitic acid are first separated using Lipidex-5000 (the peptides elute at about 30% of the first column volume, whereas palmitic acid elutes after about one column volume). After this separation, dipalmitoylated peptide is separated from mono- and nonpalmitoylated peptides by RP-HPLC. Analysis of the secondary structure by CD spectroscopy of a peptide corresponding to positions 1–17 of SP-C with and without palmitoyl groups indicates that the palmitoyl chains induce the formation of helical structure [27].

There is no detectable consensus sequence surrounding the sites of thioester-linked fatty acylation in proteins, and no enzyme responsible for the acyl-transfer reaction has been purified to date. This raises the possibility that protein fatty acylation occurs in a nonenzymatic manner. Recently, it has also been found that a synthetic peptide corresponding to the fatty acylation site of c-Yes protein tyrosine kinase undergoes nonenzymatic acylation upon incubation in the presence of lipid vesicles containing palmitoyl-coenzyme A, whereas the same peptide with a Cys → Ser replacement was not acylated [28]. Combined, these observations suggest that fatty acylation of proteins, at least in part, is mediated without catalysis and just a consequence of a close proximity of acyl-donating lipid species and Cys residues in membrane-associated proteins.

4. Structural studies of lipopeptides

Structural analysis of lipopeptides is complicated by several factors. The analysis requires the use of solvents in which the lipopeptides retain their native conformation. Organic solvents, particularly trifluoroethanol or other fluorinated alcohols, have been widely used for analysis of secondary structures of membrane-interacting peptides. The conformation of SP-C in chloroform/methanol/0.1 M HCl, 32:64:5 (by vol), was determined by two-dimensional ^{1}H NMR spectroscopy [29]. The structure reveals one very regular α-helix with nearly ideal helix geometry encompassing positions 9–34 (Fig. 3). The N-terminal eight residues, including the two palmitoylcysteine residues at positions 5 and 6, are flexibly disordered in this acidic, aqueous organic solvent. With a few exceptions the overall helical content of SP-C in different environments and determined with several different techniques is in reasonable agreement with the helical content determined in solution by NMR [30]. Early studies of FTIR spectroscopy showed that SP-C is oriented in a transmembrane manner in phospholipid bilayers [20, 31]. The size of the SP-C α-helix in organic solvents strongly supports a transmembranous orientation in phospholipid bilayers [29]. The length of the helix encompassing residues 9–34 is 37 Å and the poly-valyl part measures 23 Å. This is in very good agreement with the thickness of a fluid bilayer composed of dipalmitoylphosphatidylcholine (37 Å) and its acyl-chain part (26 Å). In contrast to the situation with the helical part, the disordered conformation of the SP-C N-terminal eight residues in solution does not likely represent the structure in a phospholipid environment. In phospholipids the N-terminal octapeptide segment is presumably anchored close to the phospholipid layer *via* the thioester-linked palmitoyl moieties.

Another problem in structural studies is that removal of lipopeptides from their natural membrane environment may cause them to denature and aggregate. This is apparently not exclusively caused by the hydrophobicity of lipopeptides. SP-C, which contains a poly-Val stretch (Fig. 4), transforms, when solubilized in organic solvents, from α-helix to β-sheet conformation with time [25]. This is compatible with the fact that Val is overrepresented in extended conformations but forms α-helixes in a lipid environment [32]. However, an SP-C analogue with all Val residues replaced with Leu is stable long-term in organic solvents and can be renatured after acid-induced denaturation [11]. Thus, the stability properties of lipopeptides cannot directly be predicted from their structural properties alone, and identification of optimal conditions for solubilization of lipopeptides remains a predominantly empirical process.

5. Analysis of recombinant proSP-B

SP-B, amoebapores (which are pore-forming polypeptides from *Entamoeba histolytica*), parts of acid sphingomyelinase, and acyloxyacylhydrolase, saposins (which promote enzymatic degradation of sphingolipids in lysosomes), and NK-lysin (which is an antibacterial and tumourolytic polypeptide from natural killer cells) constitute the saposin family of homologous polypeptides [33]. Interestingly, only SP-B of the saposins is a lipopeptide. Furthermore, the 42-kDa precursor of surfactant protein B (proSP-B) has been proposed to contain three tandem saposin repeats [34], where mature SP-B corresponds to the second of these repeats. Of these three domains, only SP-B has been isolated. In order to gain insight into the structure/function relationships of the saposins, we have expressed proSP-B in *Escherichia coli* [35].

ProSP-B, when fused to an N-terminal His-tag, forms inclusion bodies, which can be solubilized with SDS and purified to homogeneity by affinity chromatography after removal of the SDS. Purified recombinant SP-B (rproSP-B) is soluble in sodium phosphate buffer and exhibits about 35%

α-helical structure, which is similar to the approximatly 45% helix of SP-B in dodecylphosphocholine micelles estimated by CD spectroscopy. Limited proteolysis of rproSP-B occurs predominantly between the three proposed tandem saposin domains, which supports the possibility that pro-SP-B contains, in addition to SP-B, two further saposin domains [34]. This suggests that the lipopeptide SP-B does not possess unique structural properties but can be formed from polar saposins by substitution of polar residues with nonpolar ones, still preserving the polypeptide backbone fold.

6. Conclusions

Lipopeptides are important constituents of a large number of biological systems, including, for example, pulmonary surfactant, the respiratory chain, neural tissue and host-defense systems. Biochemical studies of lipopeptides are hampered by their physicochemical properties, which make them complicated to handle experimentally. However, current methodology for isolation and analysis of lipopeptides allows acquisition of high-resolution data. Challenges for the future include understanding the structural and functional importance of thioester-linked fatty acyl chains and determination of lipopeptide structures in nativelike membrane environments.

Acknowledgments
This study was supported by the Swedish Medical Research Council (project 13X-10371).

References

1 Folch J, Lees M (1951) Proteolipids: a new type of tissue lipoprotein. *J Biol Chem* 191: 807–817
2 Johansson J, Curstedt T (1997) Molecular structures and interactions of pulmonary surfactant components. *Eur J Biochem* 244: 675–693
3 Curstedt T, Jörnvall H, Robertson B, Bergman T, Berggren P (1987) Two hydrophobic low-molecular-mass protein fractions of pulmonary surfactant. Characterization and biophysical activity. *Eur J Biochem* 168: 255–262
4 Johansson J, Gröndal S, Sjövall J, Jörnvall H, Curstedt T (1992) Identification of hydrophobic fragments of α_1-antitrypsin and C1 protease inhibitor in human bile, plasma and spleen. *FEBS Lett* 299: 146–148
5 Casarett-Bruce M, Camner P, Curstedt T (1981) Changes in pulmonary lipid composition of rabbits exposed to nickel dust. *Environ Res* 26: 353–362
6 Bizzozero O, Besio-moreno M, Pasquini JM, Soto EF, Gomez CJ (1982) Rapid purification of proteolipids from rat brain subcellular fractions by chromatography on a lipophilic dextran gel. *J Chromatogr* 227: 33–44
7 Bergman T, Carlquist M, Jörnvall H (1986) Amino acid analysis by high performance liqid chromatography of phenylthiocarbamyl derivatives. In: B Wittmann-Liebold, J Salnikov, VA Erdmann (eds): *Advanced methods in protein microsequence analysis*. Berlin, 45–55
8 Stark M, Wang Y, Danielsson O, Jörnvall H, Johansson J (1998) Determination of proteins, phosphatidylethanolamine and phosphatidylserine in organic solvent extracts of tissue material by analysis of phenylthiocarbamyl derivatives. *Anal Biochem* 265: 97–102

9 Gustafsson M, Curstedt T, Jörnvall H (1997) Reverse-phase HPLC of the hydrophobic pulmonary surfactant proteins. Detection of an isoform containing N^{ϵ}-palmitoylysine. *Biochem J* 326: 799–806

10 Gustafsson M, Vandenbussche G, Curstedt T, Ruysschaert J-M, Johansson J (1996) The 21-residue surfactant peptide (LysLeu$_4$)$_4$Lys (KL$_4$) is a transmembrane α-helix with a mixed nonpolar/polar surface. *FEBS Lett* 384: 185–188

11 Nilsson G, Gustafsson M, Vandenbussche G, Veldhuizen E, Griffiths W, Sjövall J, Haagsman H, Ruysschaert J-M, Robertson B, Curstedt T et al (1998) Synthetic peptide-containing surfactants. Evaluation of transmembrane *versus* amphipathic helices and surfactant protein C poly-valyl to poly-leucyl substitution. *Eur J Biochem* 255: 116–124

12 Blondelle SE, Ostresh JM, Houghten RA, Pérez-Payá E (1995) Induced conformational states of amphipathic peptides in aqueous/lipid environments. *Biophys J* 68: 351–359

13 Griffiths WJ, Gustafsson M, Yang Y, Sjövall J, Curstedt T, Johansson J (1998) Analysis of porcine surfactant protein-C variants by nano-electrospray mass spectrometry. *Rapid Comm Mass Spectrom* 12: 1104–1114

14 Casey PJ, Buss JE (eds) (1995) Lipid modifications of proteins. *Methods Enzymol* 250

15 Hjertman M, Wejde J, Dricu A, Carlberg M, Griffiths WJ, Sjövall J, Larsson O (1997) Evidence for protein dolichylation. *FEBS Lett* 416: 235–238

16 Magee AI, Courtneidge SA (1985) Two classes of fatty acylated proteins exist in eukaryotic cells. *EMBO J* 4: 1137–1144

17 Weimbs T, Stoffel W (1992) Proteolipid protein (PLP) of CNS myelin: position of free, disulfide-bonded and fatty acid thioester-linked cysteine residues and implications for the membrane topology of PLP. *Biochemistry* 31: 12289–12296

18 Curstedt T, Johansson J, Persson P, Eklund A, Robertson B, Löwenadler B, Jörnvall H (1990) Hydrophobic surfactant-associated polypeptides: SP-C is a lipopeptide with two palmitoylated cysteine residues, where SP-B lacks convalently linked fatty acyl groups. *Proc Natl Acad Sci USA* 87: 2985–2989

19 Johansson J, Persson P, Löwenadler B, Robertson B, Jörnvall H, Curstedt T (1991) Canine hydrophobic surfactant polypeptide SP-C: a lipopeptide with one thioester-linked palmitoyl group. *FEBS Lett* 281: 119–122

20 Vandenbussche G, Clercx A, Curstedt T, Johansson J, Jörnvall H, Ruysschaert J-M (1992) Structure and orientation of the surfactant-associated protein C in a lipid bilayer. *Eur J Biochem* 203: 201–209

21 Johansson J, Nilsson G, Strömberg R, Robertson B, Jörnvall H, Curstedt T (1995) Secondary structure and biophysical activity of synthetic analogues of the pulmonary surfactant polypeptide SP-C. *Biochem J* 307: 535–541

22 Wang Z, Gurel O, Baatz JE, Notter RH (1996) Acylation of pulmonary surfactant protein-C is required for its optimal surface active interactions with phospholipids. *J Biol Chem* 271: 19104–19109

23 Qanbar R, Cheng S, Possmayer F, Schürch S (1996) Role of the palmitoylation of surfactant-associated protein C in surfactant film formation and stability. *Am J Physiol* 271: L572–L580

24 Voss T, Schäfer KP, Nielsen PF, Schäfer A, Maier C, Hannappel E, Maassen J, Landis B, Klemm B, Przybylski M (1992) Primary structure differences of human surfactant-associated proteins isolated from normal and proteinosis lung. *Biochim Biophys Acta* 1138: 261–267

25 Szyperski T, Vandenbussche G, Curstedt T, Ruysschaert J-M, Wüthrich K, Johansson J (1998) Pulmonary surfactant-associated polypeptide C in a mixed organic solvent transforms from a monomeric α-helical state into insoluble β-sheet aggregates. *Protein Sci* 7: 2533–2540

26 Creuwels LAJM, Demel RA, van Golde LMG, Benson BJ, Haagsman H (1993) Effect of acylation on structure and function of surfactant protein C at the air-liquid interface. *J Biol Chem* 268: 26752–26758

27 Salakdeh-Yousefi E, Johansson J, Strömberg R (1999) A method for *S*- and *O*-palmitoylation of peptides. Synthesis of pulmonary surfactant protein-C models; *Biochem J* 343: 557–562

28 Baño MC, Jackson CS, Magee AI (1998) Pseudo-enzymatic S-acylation of a myristoylated Yes protein tyrosine kinase peptide *in vitro* may reflect non-enzymatic S-acylation *in vivo*. *Biochem J* 330: 723–731

29 Johansson J, Szyperski T, Curstedt T, Wüthrich K (1994) The NMR structure of the pulmo-
    nary surfactant-associated polypeptide SP-C in an apolar solvent contains a valyl-rich
    α-helix. *Biochemistry* 33: 6015–6023
30 Vandenbussche G, Johansson J, Clercx A, Curstedt T, Ruysschaert J-M (1995) Structure and
    orientation of hydrophobic surfactant-associated proteins in a lipid environment. In: P
    Jollès, H Jörnvall (eds): *Interface between chemistry and biochemistry*. Birkhäuser, Basel,
    27–47
31 Pastrana B, Mautone AJ, Mendelsohn R (1991) Fourier transform infrared studies of
    secondary structure and orientation of pulmonary surfactant SP-C and its effect on the
    dynamic surface properties of phospholipids. *Biochemistry* 30: 10058–10064
32 Li SC, Deber CM (1994) A measure of helical propensity for amino acids in membrane
    environments. *Nature Struct Biol* 1: 368–373
33 Andersson M, Curstedt T, Jörnvall H, Johansson J (1995) An amphipathic helical motif
    common to tumourolytic polypeptide NK-lysin and pulmonary surfactant polypeptide
    SP-B. *FEBS Lett* 362: 328–332
34 Patthy L (1991) Homology of the precursor of the pulmonary surfactant-associated SP-B
    with prosaposin and sulfated glycoprotein-1. *J Biol Chem* 266: 6035–6037
35 Zaltash S, Johansson J (1998) Secondary structure and limited proteolysis give experimen-
    tal evidence that the precursor of surfactant protein B contains three saposin-like domains.
    FEBS Lett 423: 1–4

Proteomics in Functional Genomics
ed. by P. Jollès and H. Jörnvall

Phosphopeptide analysis

Manfredo Quadroni[1] and Peter James[2]

[1] Biomedical Research Center, University of British Columbia, 2222 Health Sciences Mall, Vancouver, B.C., Canada V6T 1Z3
[2] Protein Chemistry Laboratory, Institute for Biochemistry, Universitätsstrasse 16, ETH-Zentrum, CH-8092 Zürich, Switzerland

Summary. In this chapter we review the various methods available to the experimenter to analyse phosphorylated peptides. The initial steps in such an analysis involve the isolation of the phosphopeptides for analysis, and we outline the various current methods such as immobilised metal affinity chromatography, anti-phosphoamino acid antibodies as well as HPLC (High Pressure Liquid Chromatography) and TLC (Thin Layer Chromatography). The isolated peptides can be analysed by chemical modification followed by Edman degradation or by mass spectrometry (MS). We focus on MS methods and give examples illustrating the selective detection and sequencing of phosphopeptides.

Introduction

Protein phosphorylation has emerged, since its discovery over 4 decades ago [1], as the main mechanism by which cells modulate enzyme activity and protein-protein interactions. It is estimated that as much as one-third of the proteins expressed in a typical mammalian cell may be phosphorylated at some stage during the life of the cell [2]. Virtually all aspects of a cell's activities appear to be regulated by phosphorylation: (i) cell proliferation and differentiation, (ii) cell survival/programmed cell death, (iii) cell cycle progression, (iv) cell shape and adhesion, (v) protein secretion, (vi) endocytosis and phagocytosis, and (vii) chemotactic and sensory events.

The study of protein kinases, that is the enzymes that are responsible for protein phosphorylation has revealed an astonishing number of genes and proteins: more kinase genes are known than for any other family of proteins, all originating from a presumably common ancestor [3]. Dephosphorylating enzymes, protein phosphatases, have more recently become the focus of attention since they are obligate antagonists of the kinases in regulating an extremely dynamic equilibrium that is vital for the life of any eukaryotic cell [4]. Phosphorylation of serine, threonine and tyrosine are by far the most common regulatory events in eukaryotic cells, with a few other residues (notably Asp and Cys) only characterised as intermediates in phosphat transfer reactions. Histidine phosphate appears to play the dominant role in prokaryotic systems, though now serine and threonine phosphorylation and associated kinases have been found in bacteria [5].

Phosphorylation can modulate the function of a protein in several ways, either directly or by creating a new binding site for another protein that can act as inhibitor or activator, or by targeting a protein to a specific intra-cellular compartment. It is obviously of prime importance to determine the type and site of phosphorylation on a protein, to obtain clues as to which kinase may be responsible for the modification and what the functional significance might be. In general, the analysis of phosphorylation sites of a protein requires four steps: (i) isolation of the phosphoprotein, (ii) pro-teolytic digestion with a specific protease; (iii) analysis of the digest and identification of the phosphorylated peptides, and (iv) analysis of the phos-phopeptide(s) to precisely locate the modified amino acid.

1. Selective isolation of phosphoproteins/phosphopeptides

There is to date no straightforward, general method to purify phospho-proteins. Radiolabelling proteins using ^{32}P phosphate and then isolating the protein(s) of interest by standard means has been the method of choice for determining phosphorylation sites on proteins. Possibly one of the most powerful methods to analyse protein phosphorylation in a global manner is by two-dimensional (2D) gel electrophoresis. By radiolabelling proteins with ^{32}P phosphate and comparing the radioactivity patterns of 2D gels before and after stimulation, the proteins involved in a transduction path-way can be mapped [6, 7]. Now that ultrasensitive methods of peptide analysis have been developed, proteins can even be isolated from 1D or 2D SDS/PAGE gels and the phosphorylation sites determined.

1.1. Separating phosphopeptides from nonphosphopeptides

In order to be analysed by chemical means, phosphopeptides should be isolated as a single pure species to avoid problems in interpreting the results of sequence analysis. Phosphopeptides are notoriously difficult to handle since they have a great tendency to bind to metal surfaces such as HPLC (High Pressure Liquid Chromatography) column frits and stainless teel tubing. Separating a phosphopeptide from its nonphosphorylated analog can be readily achieved by using a counter-ion such as heptafluoro-butyric acid in a standard HPLC separation on regular reversed-phase columns in water/acetonitrile gradients [8]. A more involved method has been suggested using reversed-phase gradients together with POROS per-fusion chromatography at alkaline pH [9]. In general, separating phospho-peptides from their nonphosphorylated counterparts by HPLC is not a major problem, since phosphopeptides almost always ex-hibit a greater hydrophilicity and shorter retention times. However, it is much more diffi-cult to isolate specific phosphopeptides when the are minor components of a complex peptide mixture such as protein digest.

1.2. Extraction of phosphopeptides from complex peptide mixtures

The best general method for the specific isolation of phosphopeptides is immobilized metal affinity chromatography (IMAC). The technique was pioneered by J. Pohrat [10] and is based on the affinity of phosphoamino-acids for ferric ions bound to an iminodiacetic acid matrix. It has been refined by the groups of J. Stults [11] and R. Aebersold [12] and applied to the isolation of phosphopeptides from a protein digest, followed by the characterisation of the phosphopeptide by electrospray mass spectrometry. Miniaturisation of the affinity column is of course essential when analysing minute amounts of protein. Although effective with a purified sample, isolation of phosphopeptides or phosphoproteins by IMAC is very suscep-tible to detergents and other buffer components, and shows an appreciable degree of nonspecificity. Indeed, nonphosphorylated acidic or histidine-rich peptides are often recovered together with phosphorylated peptides. A recent report [13] indicates that some of these problems of nonspeci-ficity can be solved by using a matrix based on nitrilotriacetic acid instead of iminodiacetic acid.

1.3. Anti-phosphoaminoacid monoclonal antibodies

The availability of anti-phosphotyrosine monoclonal antibodies has con-tributed significantly to the research on signal transduction during the last decade [14]. Although their reactivity to different proteins may vary con-siderably, these antibodies have a high enough affinity to be able to detect minute amounts of tyrosine-phosphorylated proteins on Western blots and can recognise an exposed P-Tyr residue more or less independently from the nature of the surrounding amino acid sequence. Anti-PTyr antibodies have been used to sequence signal transduction proteins [15]. Unfortunate-ly, no equivalent antibody has been successfully developed so far that can recognise P-Ser and P-Thr in a sequence-independent manner, although there are commercially available antibodies that recognise with high affinity Ser/Thr-phosphorylated sequences on specific proteins with little cross-reactivity to the unphosphorylated form. As a general rule, all these antibodies behave quite poorly as affinity reagents, especially bad toward small peptides, and their main application remains in Western blotting. Attempts by one of the authors (M.Q.) to use anti-phosphotyrosine anti-bodies to purify phosphotyrosine peptides from complex mixtures were unsuccessful.

2. Phosphopeptide analysis

2.1. Classical techniques

2.1.1. Thin-layer chromatography/electrophoresis
In the past, researchers have relied heavily on radiolabelling with ^{32}P phosphate and thin-layer chromatography (TLC) or HPLC for peptide mapping and Edman degradation for the identification of the phosphorylation sites. Some of the techniques used such as the mapping of phospho-amino acids (after total acid hydrolysis of a protein) and phosphopeptide mapping by 2D TLC/thin-layer electrophoresis [16] are still of great use. The technique is highly sensitive, since a few hundred disintegrations are sufficient to generate a signal on a film on long exposure. Its main utility still lies in determining the type (Ser/Thr/Tyr) of phosphorylation, since it can provide only indirect evidence about the identity of a phosphopeptide. It can, however, be coupled with Edman degradation (albeit much more material is needed), or, more interestingly, the peptide can be recovered from the TLC plate and analysed by mass spectrometry [17].

2.1.2. Edman degradation and chemical modifications
The analysis of phosphorylation sites by Edman degradation is quite cumbersome. Three methods are commonly used. First, the peptides can be attached to a solid phase support, and then half of the eluate from the reaction cartridge after cleavage of every amino acid is delivered to a fraction collector for scintillation counting whilst the other half goes to the conversion flask for PTH (Phenylthiohydantoin) conversion and identification by HPLC (of all the amino acids except the phosphorylated ones, which cannot be identified). Second, the peptide can be spotted onto a support, and a piece removed after each cycle for radioactivity counting. Finally, an alternative method is to chemically derivatise phospho-amino acids before Edman sequence analysis. A common approach to derivatisation of phospho-amino acids is based on their susceptibility to β-elimination and subsequent reaction with ethanethiol to produce S-ethylcysteine. The derivative is now more hydrophobic, allowing the peptide to be separated from the phosphopeptide, and more important, the amino acid derivative can now be detected by standard HPLC gradients used for PTH analysis (for a review, see [18]). More recently, Fadden and Haystead [19] used ethanedithiol and 6-iodoacetamidofluorescein to generate a fluorophore-labelled amino acid. The technique is optimised only for PSer and was ineffective for PThr. Nevertheless, it has the potential to be a useful tool to specifically detect phosphopeptides and phosphorylated proteins, exploiting the high sensitivity provided by laser-induced fluorescence detection. One can imagine that a 2D gel with phosphoproteins derivatised in this way can be scanned with a fluoroimager device to provide high sensitivity of detection. More uses of chemical modification of phosphopeptides coupled to analysis by mass spectrometry (MS) will be discussed below.

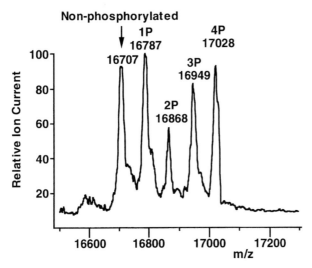

Figure 1. Deconvoluted electrospray mass spectra of intact phosphorylated calmodulin. The protein solution was infused into the electrospray source of a Finnigan MAT 700 triple quadrupole mass spectrometer. The mass range was scanned from 1000 to 2000 mass units in 2 s. The scans were averaged and the spectrum deconvoluted to give the calculated spectrum for the singly charged proteins using a data package supplied by Finnigan MAT.

2.2. Phosphopeptide analysis by MS

Recently, the development of new ionisation methods for the analysis of proteins and peptides by MS has revolutionised the field, and new methods for the analysis of phosphorylation sites have been developed. The ability to measure the masses of intact proteins is of great use. Calmodulin is phosphorylated by casein kinase II, and measurements by several groups showed that ~2 mol of phosphate was incorporated per mole of protein. Mass spectrometric analysis of the phosphorylation product (Fig. 1) revealed that this was not the whole story: phosphorylated forms carrying one, two, three or four phosphates per molecule formed about half of the sample, whereas the remaining half was composed of nonmodified molecules. A similar situation was found with phosphorylation of calmodulin by the tyrosine kinase c-fyn: incorporation of radiactive phosphate indicated a stoichiometry of one phosphate per molecule, but MS indicated the presence of 50% nonphosphorylated and 50% doubly phosphorylated forms of the protein. Similar details would have been overlooked by analysis using classical tools.

2.2.1. Identification of phosphorylated peptides by peptide mass mapping
Peptides show an increase in mass of 80 a.m.u. for each phosphate added; thus the simplest method of identifying the peptide(s) containing phosphorylation sites is mass mapping. A specific proteolytic digest is carried

out, and the masses of the peptides are determined. When the sequence of the protein is known, the phosphorylated peptides can be picked out simply by finding which predicted masses have been shifted by addition of a multiple of 80 a.m.u. [20]. If the sequence is unknown, the phosphopeptides can be picked out by comparing the digest of the phosphorylated with the nonphosphorylated sample (usually obtained by alkaline phosphatase treatment), and phosphopeptides can be identified based on the mass shift of -80 a.m.u. after dephosphorylation [21]. This method as usually been carried out by matrix-assisted laser desorption ionisation time-of-flight MS (MALDI-TOF-MS), since it is very sensitive and produces single spectra covering the entire digest mass range, dominated by singly charged ion species, making it the technique of choice for extensive peptide mapping. Recently an analogous method for the determination of tyrosinephosphorylated peptides was demonstrated by Aebersold's group [22] using a triple quadrupole mass spectrometer. Half of the protein digest is passed through an on-line enzyme reactor containing alkaline phosphatase onto a reverse-phase HPLC column, and the peptides are analysed by electrospray ionisation MS. The other half is analysed without dephosphorylation, and a comparison of the two mass profiles indicates the phosphorylated peptides.

The main limitation of differential peptide mapping is that the phosphorylation site is defined as being on a certain peptide, and the phosphorylated residue is not defined. If the peptide contains more than one possible phosphorylation site, either further digestions will be necessary to divide the sites and allow identification, or the peptide phosphorylation site must be determined by fragmentation in the mass spectrometer.

2.2.2. MS/MS fragmentation (daughter ion scanning)

Peptides can be fragmented in a mass spectrometer, and the sequence and site(s) of phosphorylation determined. In addition to amide backbone bond fragmentation giving rise to "sequence ions", phosphopeptides show side chain losses of H_3PO_4 or HPO_3. Triple quadrupole mass spectrometers can be programmed to analyse the masses of all the peptides entering the spectrometer (normal scan mode, Fig. 2a) and then to switch to daughter ion scanning (MS/MS) mode (Fig. 2b) to accumulate sequence data. Spectra of a serine-phosphorylated peptide (Fig. 3a), and a closely sequence-related tyrosine-phosphorylated peptide (Fig. 3b) obtained under almost identical conditions are given for comparison. The serine (and this also applies to threonine) loses the phosphate group very easily by β-elimination mostly as H_3PO_4, whereas the tyrosine preferentially losses phosphate as HPO_3. Annan and Carr [23] have shown that it is also possible to distinguish Ser/Thr from Tyr phosphorylation in post-source decay (PSD) spectra of phosphopeptides obtained by MALDI-TOF-MS. They demonstrated that in the case of Ser/Thr peptides, the ion with mass [M-98] is more intense than the one with [M-80], whereas the opposite is true for peptides phosphorylated on tyrosine. In some cases, if the consensus sequence prefer-

(a) NORMAL SCANNING MODE

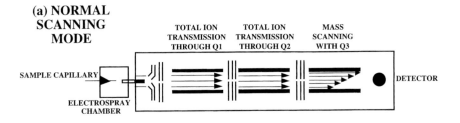

(b) DAUGHTER SCANNING MODE

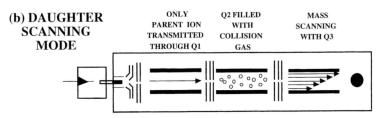

Figure 2. Schematic diagrams of (a) normal an (b) daughter ion (MS/MS) scanning. A triple quadrupole mass spectrometer may be thought of as three mass filters in series. In normal scanning mode the first two filters are set to allow all ions to pass, and the spectrum is accumulated by scanning a fixed-width window over the mass range using the third filter. The ions passing through the window are detected and the mass determined by comparison of the RF and DC values at that point with a calibration table. In MS/MS mode the first filter is used to select a mass window around an ion allowing it to pass and removing all others. The ion is accelerated into the second quadrupole into the collision gas (usually argon) where it undergoes multiple collisions, causing fragmentation. The daughter ions are analysed by scanning the third filter.

ences of the kinase are known as well the mass of the phosphorylated tryptic peptide, this is enough to unequivocally assign the modification to a specific residue.

2.2.3. Neutral loss scanning

Daughter ion spectra of peptides that contain phosphoserine and phosphothreonine side chains show that these side chains have a strong tendency to undergo β-elimination, losing H_3PO_4 (Δm -98). A triple-stage quadrupole mass spectrometer operating in the neutral loss mode (Fig. 4a) can be set up to detect only those ions that lose a fragment of a particular mass (in this case 98 a.m.u.) after passage in the collision chamber. Since peptides can assume multiple charges after electrospray ionisation, the difference in m/z observed for each loss of a phosphate group will be 98 for a 1^+ ion and 49 for a 2^+ ion. This simplifies the identification of phosphopeptides, but a second run must be made to find the tyrosine-phosphorylated peptides (losing 80). However, neutral loss measurements are inherently less sensitive than parent ion scanning and subject to "contamination" from the signals generated by other side chains (i.e. b1 ions of Pro and Val, with

(a)

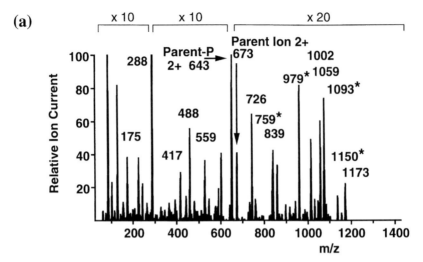

b1	b2	b3	b4					P				
116	173	287	344									

ASP - GLY - ASN - GLY - TYR - ILE - SER - ALA - ALA - GLU - Lxx - ARG

	1230	1173	1059	1002	839	726	559	488	417	288	175
	y11	y10	y9	y8	y7	y6	y5	y4	y3	y2	y1
-P	1150	1093	979	922	759						

(b)

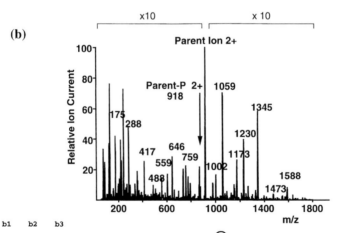

b1	b2	b3								P						
100	247	362														

VAL - PHE - ASP - LYS - ASP - GLY - ASN - GLY - TYR - ILE - SER - ALA - ALA - GLU - Lxx - ARG

			1588	1473	1345	1230	1173	1059	1002	759	646	559	488	417	288	175
				y11	y10	y9	y8	y7	y6	y5	y4	y3	y2	y1		

masses of, respectively, 97 and 99 a.m.u.), and it is sometimes difficult to distinguish between a loss of two phosphates from a 2^+ and loss of one phosphate from a 1^+ ion. Neutral loss analysis of digests by LC-MS (Liquid Chromatography-Mass Spectrometry) has been employed in several studies [24, 25] but appears to be limited to cases where a sufficient amount of material is available.

2.2.4. Parent ion scanning

Methods based on parent ion scanning (Fig. 4b) try of specifically detect phosphopeptides by measuring the signal generated by the released phosphate group itself. The protonated form of the phosphoric acid group is not normally observed, even in acidic conditions. The deprotonated PO_3^- ion (m/z 79) has proven to be the best diagnostic ion for specifically detecting phosphopeptides, since it is rarely observed as a product of fragmentation of nonphosphorylated peptides (though sometimes glycopeptides can give this signal). This type of approach applied to the analysis of complex mixtures has become very attractive recently, after the development of ultra low flow spraying sources made popular by the nanoelectrospray ion source. Selective detection and sequencing of phosphopeptides at the femtomole level using parent ion scanning of m/z −79 has been shown [26–28]. Besides the benefits derived from increased efficiency and sensitivity, this off-line method allows one to change the pH of the analyte; this is of great advantage when working with phosphopeptides, making it possible to analyse samples at pH >7 and thus to obtain the maximum signal for the −79 ion. The conditions can then be easily reverted to acidic to perform MS/MS analysis on the candidate ions found to give signals in the scan for parents of −79.

A normal scan spectrum of a digest of a phosphorylated protein is shown in Figure 5a together with the negative mode spectrum obtained by parent ion scanning, Figure 5b. The phosphopeptide is easy to pick out, and the daughter ion (MS/MS) spectrum performed on the doubly charged species is shown in Figure 6. The main advantages of this technique include (i) sensitivity: phosphopeptides can be detected that are minor components of the mixture, (ii) specificity: very little "cross-reactivity" observed by non-

---

Figure 3. Daughter ion (MS/MS) mass spectra MS/MS of a Ser/Thr-phosphorylated peptide and (b) a closely related Tyr-phosphorylated peptide. (a) The phosphopeptide was subjected to collisionally activated dissociation using 2.5 m Torr of argon and a collision energy of 20 eV. The sequence is shown below. The superscript* indicates the loss of (PO_3) and 2^+ that the ion is doubly charged (all others are singly charged). Areas blown up by magnification factors are indicated above the spectrum by a factor x. Ions are named according to Roepstorff and Fohlman [35]. The spectrum shows many ions which readily lose phosphate as well as water. (b) The phosphopeptide was subjected to collisionally activated dissociation using 2.5 mTorr of argon and a collision energy of 20 eV. The spectrum was recorded by selecting the $MH2^+$ parent ion (m/z 918) under the same conditions of resolution, collision energy and gas pressure as in (a) The loss of phosphate from the parent ion and the daughters is much less pronounced than with the Ser(P)-containing peptide, and loss of 98 is not observed.

(a) PARENT SCANNING MODE

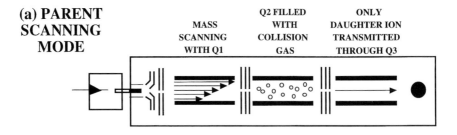

(b) NEUTRAL LOSS SCANNING MODE

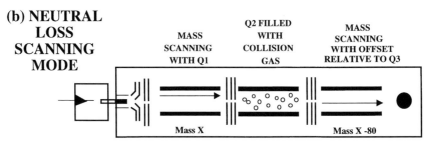

Figure 4. Schematic diagrams of (a) parent and (b) neutral loss scanning. (A) The first mass filter is set to scan in normal mode. As peptide of defined mass pass the filter, they are subject to collisionally induced dissociation in the second quadrupole, and the third filter is set to detect a single mass only. The mass window can be set to ions diagnostic of peptides such as 86 for isoleucine and leucine or 147 for lysine, allowing peptides to be picked out even when their intensity is the same as the chemical noise in the system. (B) Neutral loss scanning sets the first and third filter scanning in parallel but offset by the difference in mass between phospho- and nonphosphopeptides (80 for 1^+, 40 for 2^+ ions).

Table 1. Result of SEQUEST search using the MS/MS spectrum from Figure 6. The molecular mass of tyrosine was modified to account for phosphorylation

SEQUEST OUTPUT
Molecular Biotechnology, Univ. of Washington, J. Eng/J. Yates
Mass = 2283.0 (+2), $\underline{Y = 243.18,}$ Enzyme: Trypsin, OWL database.

#	Rank/Sp	(M+H)+	Cn	Ions	Reference	Peptide
1.	1/1	2283.3	1.0000	15/36	pir | P02730 |	(K) ATFDEEEGRDEY*DEVAMPV
2.	2/124	2282.2	0.8477	10/36	pir | U50719 |	(K) KDTGNYGCNATSSIGYVYK
3.	3/8	2281.7	0.8460	13/44	pir | X95938 |	(R) AAADEGEIPIGAVIVKGQIVAR
4.	4/2	2282.5	0.8071	12/40	pir | U00046 |	(K) SQEATSFSSIALGGDEWVLKR
5.	5/35	2284.4	0.7819	11/38	pir | S58530 |	(R) LTLCSYGEGGNGFQCPTGYR

1. pir | P02730 | B3AT_HUMAN BAND 3 ANION TRANSPORT PROTEIN
2. pir | U50719 | MSU50719 MSU 50719 NID: g1708634 – tobacco hornworm.
3. pir | X95938 | PGPGAAGEN1 NID: g1296967 – *Porphyromonas gingivalis.*
4. pir | U00046 | CELR13F6 CELR13F6 NID: g470358 – *Caenorhabditis elegans*
5. pir | S58530 | S58530 elicitor NIP1 precursor – *Rhynchosporium secalis*

(a) Normal scan in positive mode

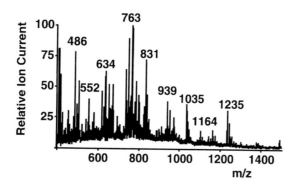

(b) Parent ion scan in negative mode

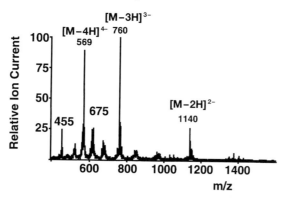

Figure 5. Negative mode parent scanning specifically detects a tyrosine-phosphorylated peptide in a crude mixture. A tryptic digest of a protein was analyzed unfractionated by nano-electrospray mass spectrometry. Panel (a) shows the full scan of the mixture in positive mode under acidic conditions with the masses of some of the most intense ions labelled. The same peptide mixture was brought to pH 10 by the addition of ammonia and analyzed in negative mode (b), with the instrument set to detect the parent ions of $m/z = 79^-$. The charge states $[M-2H]^{2-}$ to $[M-5H]^{5-}$ of a peptide with $[M + H]^+ = 2282.3$ (avg mass) are observed. The phosphopeptide was barely detectable in positive mode scan.

phosphopeptides, (iii) rapidity and ease: often a parent scan can be collected and the same aliquot of sample (1–2 µl) can be acidified and analysed in positive mode to collect MS/MS data. The MS/MS data not only allow the identification of the site of phosphoryation but can also be used in computer searches to identify the protein that is being analysed [25, 29, 30]. The results of a SEQUEST database search using the spectrum shown in Figure 6 are given in Table 1. The protein was clearly identified as erythocyte band 3, and the mass mapping data supported this conclusion.

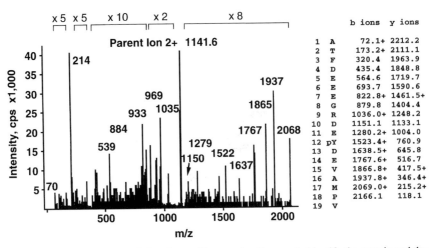

Figure 6. MS/MS analysis of phosphopeptides can simultaneously identify the protein and the site of phosphorylation. MS/MS spectrum of the phosphorylated peptide (*m/z* 2282) found in the protein digest shown in Figure 5. The doubly charged ion *m/z* = 1141.6 was used as the parent. The spectrum is labelled to show the most important sequence ions. A search of the protein sequence database OWL using the program SEQUEST identified the protein as mouse band III protein, and simultaneously the spectrum allows the phosphorylation site to be defined. The sequence of the phosphopeptide is shown, together with the ions matched in the search (labelled with +).

2.2.5. Multiply phosphorylated peptides

Numerous proteins in nature are heavily phosphorylated, and the modification can have both structural or functional significance. The analysis of these proteins to determine the exact phosphorylation sites is usually very challenging, since after digestion one is often confronted with multiply phosphorylated peptides. The extensive loss of phosphate in MS/MS experiments by these multiply phosphorylated peptides makes interpretation of daughter spectra very difficult. One such example is profilaggrin, a protein containing as much as 400 mol of phosphate/mol, and the analysis of its phosphorylation sites [24] is an elegant example of a successful strategy to tackle this kind of problem. Extensive peptide mapping by LC-MS and neutral loss experiments were used to identify a set of candidate phosphopeptides. β-Elimination by treatment with Ba(OH)$_2$ was then used to convert PSer and PThr to dehydroalanine and dehydroamino-2-butyric acid, respectively. The resulting peptides display "normal" fragmentation patterns, and the "unusual" mass of the dephosphorylated amino acids simplifies the conclusive assignment of the sites of phosphorylation. This technique would be difficult to extend to PTyr residues, and there is little need to do so since there are no reports of heavily tyrosine-phosphorylated proteins. A similar analysis of the phosphorylation sites of the human high molecular weight neurofilament protein has been carried out [25]; no chemical modification was used in this case, but computer searches of a limited sequence database using MS/MS data helped in identifying several of the phosphorylation sites.

3. Conclusions and perspectives

The application of biological MS has radically changed the detection and analysis of peptides and proteins. Today it is possible to locate phosphorylated amino acids in a polypeptide sequence with high sensitivity, speed, accuracy, and without the need to use radioactive isotopes. In the future we can expect to see improvements in MS technology that will increase sensitivity, so that one femtomole of protein will be sufficient to analyse its phosphorylation sites. The use of microfabricated devices and CZE (Capillary Zonal Electrophoresis) on-line with a mass spectrometer [31, 32] may help to push the sensitivity of detection of peptides down to the attomole range. The same technology could be easily modified to perform on-line IMAC chromatography to specifically isolated phosphopeptides. More work is required to improve sample processing upstream, minimising losses of phosphopeptides due to adsorption to metal surfaces and hydrolysis due to exposure to extreme pH (e.g. in alkaline regions of 2D gels). At the same time, the development of mass spectrometers with new architectures promises significant improvements in the sensitivity of analysis and in the quality of the sequence information that can be obtained. In particular, MALDI ion-trap instruments have already been used to successfully determine phosphorylation sites [33, 34]. One likely development in the future should be the application of multiple fragmentation steps (MS^n) to analyse multiply phosphorylated peptides in an ion-trap (cyclotron) mass spectrometer.

References

1 Krebs EG, Fischer EH (1956) The phosphorylase b toa converting enzyme of rabbit skeletal muscle. *Biochim Biophys Acta* 20: 150–157

2 Hubbard MJ, Cohen P (1993) On target with a new mechanism for the regulation of protein phosphorylation. *Trends Biochem Sci* 18: 172–177

3 Hunter T (1991) Protein kinase classification. *Methods Enzymol* 200: 3–37

4 Hunter T (1995) Protein kinases and phosphatases: the yin and yang of protein phosphorylation and signaling. *Cell* 80: 225–236

5 Galinier A, Kravanja M, Engelmann R, Hengstenberg W, Kilhoffer MC, Deutscher J, Haiech J (1998) New protein kinase and protein phosphatase families mediate signal transduction in bacterial catabolite repression. *Proc Natl Acad Sci USA* 95: 1823–1828

6 Guy GR, Philip R, Tan YH (1994) Analysis of cellular phosphoproteins by two-dimensional gel electrophoresis: applications for cell signaling in normal and cancer cells. *Electrophoresis* 15: 417–440

7 Djafarzadeh S, Niggli V (1997) Signaling pathways involved in dephosphorylation and localization of the actin-binding protein cofilin in stimulated human neutrophils. *Exp Cell Res* 236: 427–435

8 Ohguro H, Palczewski K (1995) Separation of phospho- and nonphosphopeptides using reverse phase column chromatography. *FEBS Lett* 368: 452–454

9 Matsumoto H, Kahn ES, Komori N (1997) Separation of phosphopeptides from their non-phosphorylated forms by reversed-phase POROS perfusion chromatography at alkaline pH. *Anal Biochem* 251: 116–119

10 Andersson L, Porath J (1986) Isolation of phosphoproteins by immobilized metal (Fe3+) affinity chromatography. *Anal Biochem* 154: 250–254

11 Nuwaysir LM, Stults JT (1993) Electrospray ionization mass spectrometry of phospho-peptides isolated by on-line immobilized metal-ion affinity chromatography. *J Am Soc Mass Spectrom* 4: 662–669

12 Watts JD, Affolter M, Krebs DL, Wange RL, Samelson LE, Aebersold R (1994) Identification by electrospray ionization mass spectrometry of the sites of tyrosine phosphorylation induced in activated Jurkat T cells on the protein tyrosine kinase ZAP-70. *J Biol Chem* 269: 29520–29529

13 Neville DC, Rozanas CR, Price EM, Gruis DB, Verkman AS, Townsend RR (1997) Evidence for phosphorylation of serine 753 in CFTR using a novel metal-ion affinity resin and matrix-assisted laser desorption mass spectrometry. *Protein Sci* 6: 2436–2445

14 Roberts TM, Kaplan D, Morgan W, Keller T, Mamon H, Piwnica-Worms H, Druker B, Cohen B, Schaffhausen B, Whitman M et al (1988) Tyrosine phosphorylation in signal transduction. *Cold Spring Harbor Symp Quant Biol* 53: 161–171

15 Gold MR, Yungwirth T, Sutherland CL, Ingham RJ, Vianzon D, Chiu R, van Ostveen I, Morrison HD, Aebersold R (1994) Purification and identification of tyrosine-phosphorylated proteins from B lymphocytes stimulated through the antigen receptor. *Electrophoresis* 15: 441–453

16 van der Geer P, Hunter T (1994) Phosphopeptide mapping and phosphoamino acid analysis by electrophoresis and chromatography on thin-layer cellulose plates. *Electrophoresis* 15: 544–554

17 Affolter M, Watts JD, Krebs DL, Aebersold R (1994) Evaluation of twodimensional phosphopeptide maps by electrospray ionization mass spectrometry of recovered peptides. *Anal Biochem* 223: 74–81

18 Meyer HE, Eisermann B, Heber M, Hoffmann-Posorske E, Korte H, Weight C, Wegner A, Hutton T, Donella-Deana A, Perich JW (1993) Strategies for nonradioactive methods in the localization of phosphorylated amino acids in proteins. *FASEB J* 7: 776–782

19 Fadden P, Haystead TA (1995) Quantitative and selective fluorophore labeling of phospho-serine on peptides nd proteins: characterization at the attomole level by capillary electrophoresis and laser-induced fluorescence. *Anal Biochem* 225: 81–88

20 Yip TT, Hutchens TW (1992) Mapping and sequence-specific identification of phospho-peptides in unfractionated protein digest mixtures by matrix-assisted laser desorption/ionization time-of-flight mass spectrometry. *FEBS Lett* 308: 149–153

21 Liao PC, Leykam J, Andrews PC, Gage DA, Allison J (1994) An approach to locate phosphorylation sites in a phosphoprotein: mass mapping by combining specific enzymatic degradation with matrix-assisted laser desorption/ionization mass spectrometry. *Anal Biochem* 219: 9–20

22 Amankwa LN, Harder K, Jirik F, Aebersold R (1995) High-sensitivity determination of tyrosine-phosphorylated peptides by on-line enzyme reactor and electrospray ionization mass spectrometry. *Protein Sci* 4: 113–125

23 Annan RS, Carr SA (1996) Phosphopeptide analysis by matrix-assisted laser desorption time-of-flight mass spectrometry. *Anal Chem* 68: 3413–3421

24 Resing KA, Johnson RS, Walsh KA (1995) Mass spectrometric analysis of 21 phosphorylation sites in the internal repeat of rat profilaggrin, precursor of an intermediate filament associated protein. *Biochemistry* 34: 9477–9487

25 Jaffe H, Veeranna, Shetty KT, Pant HC (1998) Characterization of the phosphorylation sites of human high molecular weight neurofilament protein by electrospray ionization tandem mass spectrometry and database searching. *Biochemistry* 37: 3931–3940

26 Carr SA, Huddleston MJ, Annan RS (1996) Selective detection and sequencing of phosphopeptides at the femtomole level by mass spectrometry. *Anal Biochem* 239: 180–192

27 Wilm M, Neubauer G, Mann M (1996) Parent ion scans of unseparated peptide mixtures. *Anal Chem* 68: 527–533

28 Cleverley KE, Betts JC, Blackstock WP, Gallo JM, Anderton BH (1998) Identification of novel *in vitro* PKA phosphorylation sites on the low and middle molecular mass neuro-filament subunits by mass spectrometry. *Biochemistry* 37: 3917–3930

29 Yates JR III, Eng JK, McCormack AL (1995) Mining genomes: correlating tandem mass spectra of modified and unmodified peptides to sequences in nucleotide databases. *Anal Chem* 67: 3202–3210

30 Mann M, Wilm M (1994) Error-tolerant identification of peptides in sequence databases by peptide sequence tags. *Anal Chem* 66: 4390–4399

31 Figeys D, Ning Y, Aebersold R (1997) A microfabricated device for rapid protein identification by microelectrospray ion trap mass spectrometry. *Anal Chem* 69: 3153–3160

32 Figeys D, Ducret A, Aebersold R (1997) Identification of proteins by capillary electrophoresis-tandem mass spectrometry. Evaluation of an on-line solid-phase extraction device. *J Chrom* A 763: 295–306

33 Jonscher KR, Yates JR III (1997) Matrix-assisted laer desorption ionization/quadrupole ion trap mass spectrometry of peptides. Application to the localization of phosphorylation sites on the P protein from Sendai virus. *J Biol Chem* 272: 1735–1741

34 Qin J, Chait BT (1997) Identification and characterization of posttranslational modifications of proteins by MALDI ion trap mass spectrometry. *Anal Chem* 69: 4002–4009

35 Roepstorff P, Fohlman J (1984) Proposal for a common nomenclature for sequence ions in mass spectra of peptides. *Biomed Environ Mass Spectrom* 11: 601–603

Proteomics in Functional Genomics
ed. by P. Jollès and H. Jörnvall
© 2000 Birkhäuser Verlag Basel/Switzerland

Bioinformatics in protein analysis

Bengt Persson

Stockholm Bioinformatic Centre and Department of Medical Biochemistry and Biophysics, Karolinska Institutet, SE-171 77 Stockholm, Sweden

Summary. The chapter gives an overview of bioinformatic techniques of importance in protein analysis. These include database searches, sequence comparisons and structural predictions. Links to useful World Wide Web (WWW) pages are given in relation to each topic.

Databases with biological information are reviewed with emphasis on databases for nucleotide sequences (EMBL, GenBank, DDBJ), genomes, amino acid sequences (Swissprot, PIR, TrEMBL, GenePept), and three-dimensional structures (PDB). Integrated user interfaces for databases (SRS and Entrez) are described. An introduction to databases of sequence patterns and protein families is also given (Prosite, Pfam, Blocks).

Furthermore, the chapter describes the widespread methods for sequence comparisons, FASTA and BLAST, and the corresponding WWW services. The techniques involving multiple sequence alignments are also reviewed: alignment creation with the Clustal programs, phylogenetic tree calculation with the Clustal or Phylip packages and tree display using Drawtree, njplot or phylo_win.

Finally, the chapter also treats the issue of structural prediction. Different methods for secondary structure predictions are described (Chou-Fasman, Garnier-Osguthorpe-Robson, Predator, PHD). Techniques for predicting membrane proteins, antigenic sites and posttranslational modifications are also reviewed.

Introduction

The number of databases providing biological information is steadily increasing. A new discipline, bioinformatics, has emerged from the growing possibilities to utilise the information in these databases to better understand biological processes and to interpret experimental data. This chapter will focus on sequence and sequence-related databases for general purposes. Apart from these, there exist a number of specialised databases focusing on a single enzyme, protein family or disease. Each year, the first issue of *Nucleic Acids Research* lists these databases. Links to biologically relevant databases can be found on the WWW (World Wide Web) pages of EBI (European Bioinformatics Institute, Hinxton, England; http:///www.ebi.ac.uk), the University of Geneva, Switzerland (http://www.expasy.ch), and NCBI (National Center for Biological Information, Bethesda, MD, USA; http://ncbi.nlm.nih.gov). They may also be found using any WWW search engine, e.g. Altavista (http://www.altavista.digital.com).

For more than 30 years, attempts have been made to predict protein properties, structure and function from the primary structure alone. One method is to compare sequences for similarities that reflect properties in common. Another method is to screen the sequence for patterns typical of a protein family, structural property or posttranslational modification. Over the years, the prediction methods have gradually improved thanks to the increasing amount of data available and improved calculation facilities due to faster computers. For instance, it is now possible to predict transmembrane segments with very high reliability (above 95%), whereas secondary structure prediction methods still achieve only up to 75% accuracy. The prediction methods are especially valuable in the studies of the complete genomes resulting in large amounts of sequence data.

1. Databases

1.1. Sequence databases

1.1.1. General properties

Databases can be classified according to the type of information: nucleotide sequences, protein sequences, three-dimensional structures, sequence patterns, bibliographic data, linkage analysis, classification systems. Most databases have a similar structure. As an example, a Swissprot entry is shown in Figure 1. Each line starts with a two-letter code, designating the type of data. For example, DE denotes description line, DT date line, and SQ sequence line. Detailed descriptions of all codes are available in the documentation for each database.

Databases are normally available for downloading via anonymous ftp (Tab. 1). Several database maintainers also offer a WWW interface. As the database sites are subject to changes, an updated version of Table 1 is available on http://www.cbb.ki.se/databases.html.

1.1.2. Nucleotide sequence databases

Nucleotide sequence data are collected into one of three internationally collaborating databases: EMBL [1] (by EBI), GenBank [2] (by NCBI), or DDBJ (the DNA Database of Japan, Mishima). Sequence data are exchanged daily between these databases. The databases are distributed via the Internet and on CD-ROM. Via the Internet, the user gets access to daily updated versions of the databases (Tab. 1).

In order to speed up searches in these databases, the databases are subdivided according to the origin of the sequences, offering the possibility to search only a subset. The present subdivisions of the EMBL database are listed in Table 2. The GenBank database is organised similarly. For convenient file transfer, the largest subgroups are divided into multiple files.

Figure 1. A Swissprot database entry. The entry shown is ADHX_HUMAN (human alcohol dehydrogenase*x*). Each line starts with a two-letter code designating the type of information of that line.

```
ID   ADHX_HUMAN      STANDARD;      PRT;   373 AA.
AC   P11766;
DT   01-OCT-1989 (REL. 12, CREATED)
DT   01-APR-1990 (REL. 14, LAST SEQUENCE UPDATE)
DT   01-NOV-1997 (REL. 35, LAST ANNOTATION UPDATE)
DE   ALCOHOL DEHYDROGENASE CLASS III CHI CHAIN (EC 1.1.1.1) (GLUTATHIONE-
DE   DEPENDENT FORMALDEHYDE DEHYDROGENASE) (EC 1.2.1.1) (FDH).
GN   ADH5 OR ADHX OR FDH.
OS   HOMO SAPIENS (HUMAN).
OC   EUKARYOTA; METAZOA; CHORDATA; VERTEBRATA; TETRAPODA; MAMMALIA;
OC   EUTHERIA; PRIMATES.
RN   [1]
RP   SEQUENCE FROM N.A.
RX   MEDLINE; 90056459.
RA   SHARMA C.P., FOX E.A., HOLMQUIST B., JOERNVALL H., VALLEE B.L.;
RL   BIOCHEM. BIOPHYS. RES. COMMUN. 164:631-637(1989).
RN   [2]
RP   SEQUENCE FROM N.A.
RX   MEDLINE; 90026418.
RA   GIRI P.R., KRUG J.F., KOZAK C., MORETTI T., O'BRIEN S.J.,
RA   SEUANEZ H.N., GOLDMAN D.;
RL   BIOCHEM. BIOPHYS. RES. COMMUN. 164:453-460(1989).
RN   [3]
RP   SEQUENCE FROM N.A.
RX   MEDLINE; 93077045.
RA   HUR M.W., EDENBERG H.J.;
RL   GENE 121:305-311(1992).
RN   [4]
RP   SEQUENCE.
RC   TISSUE=LIVER;
RX   MEDLINE; 88209465.
RA   KAISER R., HOLMQUIST B., HEMPEL J., VALLEE B.L., JOERNVALL H.;
RL   BIOCHEMISTRY 27:1132-1140(1988).
RN   [5]
RP   X-RAY CRYSTALLOGRAPHY (2.7 ANGSTROMS).
RX   MEDLINE; 97170743.
RA   YANG Z.-N., BOSRON W.F., HURLEY T.D.;
RL   .
CC   -!- FUNCTION: CLASS-III ADH IS REMARKABLY INEFFECTIVE IN OXIDIZING
CC       ETHANOL, BUT IT READILY CATALYZES THE OXIDATION OF LONG-CHAIN
CC       PRIMARY ALCOHOLS AND THE OXIDATION OF S-(HYDROXYMETHYL)
CC       GLUTATHIONE.
CC   -!- CATALYTIC ACTIVITY: ALCOHOL + NAD(+) = ALDEHYDE OR KETONE + NADH.
CC   -!- CATALYTIC ACTIVITY: FORMALDEHYDE + GLUTATHIONE + NAD(+) =
CC       S-FORMYLGLUTATHIONE + NADH.
CC   -!- COFACTOR: REQUIRES ZINC FOR ITS ACTIVITY.
CC   -!- SUBUNIT: HOMODIMER.
CC   -!- SUBCELLULAR LOCATION: CYTOPLASMIC.
CC   -!- THERE ARE 7 DIFFERENT ADH'S ISOZYMES IN HUMAN: THREE BELONGS TO
CC       CLASS-I: ALPHA, BETA, AND GAMMA, ONE TO CLASS-II: PI, ONE TO
CC       CLASS-III: CHI, ONE TO CLASS-IV: ADH7 AND ONE TO CLASS-V: ADH6.
CC   -!- SIMILARITY: BELONGS TO THE ZINC-CONTAINING ALCOHOL DEHYDROGENASE
CC       FAMILY. BELONGS TO THE ADH CLASS-III SUBFAMILY.
DR   EMBL; M30471; G178134; -.
DR   EMBL; M29872; G178132; -.
DR   EMBL; M81118; G178130; -.
DR   EMBL; M81112; G178130; JOINED.
DR   EMBL; M81113; G178130; JOINED.
DR   EMBL; M81114; G178130; JOINED.
DR   EMBL; M81115; G178130; JOINED.
DR   EMBL; M81116; G178130; JOINED.
DR   EMBL; M81117; G178130; JOINED.
DR   PIR; A33428; DEHUC2.
DR   PIR; A36739; A36739.
DR   PIR; JH0789; JH0789.
DR   PDB; 1TEH; 07-DEC-96.
DR   MIM; 103710; -.
DR   MIM; 136490; -.
DR   PROSITE; PS00059; ADH_ZINC; 1.
KW   OXIDOREDUCTASE; ZINC; NAD; MULTIGENE FAMILY; ACETYLATION;
KW   3D-STRUCTURE.
FT   INIT_MET      0      0
FT   MOD_RES       1      1       ACETYLATION.
FT   METAL        44     44       ZINC (CATALYTIC).
FT   METAL        66     66       ZINC (CATALYTIC).
FT   METAL        96     96       ZINC (SECOND ATOM).
FT   METAL        99     99       ZINC (SECOND ATOM).
FT   METAL       102    102       ZINC (SECOND ATOM).
FT   METAL       110    110       ZINC (SECOND ATOM).
FT   METAL       173    173       ZINC (CATALYTIC).
FT   CONFLICT    166    166       D -> Y (IN REF. 2).
FT   CONFLICT    245    245       F -> L (IN REF. 2).
SQ   SEQUENCE   373 AA;  39593 MW;  0D7C9710 CRC32;
     ANEVIKCKAA VAWEAGKPLS IEEIEVAPPK AHEVRIKIIA TAVCHTDAYT LSGADPEGCF
     PVILGHEGAG IVESVGEGVT KLKAGDTVIP LYIPQCGECK FCLNPKTNLC QKIRVTQGKG
     LMPDGTSRFT CKGKTILHYM GTSTFSEYTV VADISVAKID PLAPLDKVCL LGCGISTGYG
     AAVNTAKLEP GSVCAVFGLG GVGLAVIMGC KVAGASRIIG VDINKDKFAR AKEFGATECI
     NPQDFSKPIQ EVLIEMTDGG VDYSFECIGN VKVMRAALEA CHKGWGVSVV VGVAASGEEI
     ATRPFQLVTG RTWKGTAFGG WKSVESVPKL VSEYMSKKIK VDEFVTHNLS FDEINKAFEL
     MHSGKSIRTV VKI
//
```

Table 1. Links to anonymous ftp sites of different databases
(cf. http://www.cbb.ki.se/databases.html)

Database	FTP site
Protein sequences	
Swissprot	ftp://www.expasy.ch/databases/swiss-prot/
	ftp://ftp.ebi.ac.uk/pub/databases/swissprot/release/
Swissnew	ftp://www.expasy.ch/databases/swiss-prot/updates/
	ftp://ftp.ebi.ac.uk/pub/databases/swissprot/updates/
TrEMBL	ftp://ftp.ebi.ac.uk/pub/databases/trembl/
Genpept	ftp://ncbi.nlm.nih.gov/genbank/genpept.fsa.Z
Genpept, recent addtions	ftp://ncbi.nlm.nih.gov/genbank/daily/
PIR	ftp://nbrf.georgetown.edu/pir/
	ftp://ftp.ebi.ac.uk/pub/databases/pir/
NCBI's nonredundant database	ftp://ncbi.nlm.nih.gov/blast/db/nr.Z
KIND	ftp://ftp.mbb.ki.se/pub/KIND
Nucleotide sequences	
EMBL	ftp://ftp.ebi.ac.uk/pub/databases/embl/
GenBank	ftp://ncbi.nlm.nih.gov/genbank/
EST	ftp://ncbi.nlm.nih.gov/blast/db/est.Z
STS	ftp://ncbi.nlm.nih.gov/blast/db/sts.Z
Completed genomes	ftp://ftp.ebi.ac.uk/pub/databases/embl/genomes/
	ftp://ftp.tigr.org/pub/data/
Three-dimensional structures	
PDB	http://www.rcsb.gov/pdb/index.html
	ftp://ftp.ebi.ac.uk/databases/pdb/fullrelease/
Sequence patterns	
Prosite	ftp://www.expasy.ch/databases/prosite/
	ftp://ftp.ebi.ac.uk/pub/databases/prosite/
Pfam	ftp://ftp.cgr.ki.se/pub/esr/Pfam/

Table 2. Subdivisions of the EMBL database

Subdivision	Abbreviation
Bacteriophage	PHG
ESTs	EST
Fungi	FUN
High-throughput genome	HTG
Genome survey sequences	GSS
Human	HUM
Invertebrates	INV
Organelles	ORG
Other mammals	MAM
Other vertebrates	VRT
Plants	PLN
Prokaryotes	PRO
Rodents	ROD
STSs	STS
Synthetic	SYN
Unclassified	UNC
Viruses	VRL

1.1.3. Expressed sequence tags

Expressed sequence tags (ESTs) are the outcome from automated large-scale sequencing of messenger RNA (mRNA), e.g. in a collaboration between Merck & Co. and Washington University [3, 4]. These constitute a large proportion of EMBL/GenBank (72% in 1998 [2]). Even though these mRNAs are only partial and contain sequence errors, they constitute a gold mine for the researcher. By aligning multiple sequences, several sequence errors can be eliminated. The multiple sequences also give information about heterogeneity. In the EST database there is a preponderance of sequences with many copies of mRNA in the cell, whereas proteins synthesised in low amounts are weakly represented.

1.1.4. Sequence tagged sites

To obtain physical maps of genomic organisation, a number of unique "landmarks" are characterised, denoted STSs (sequence tagged sites) [5]. These can be used as reference points for organising data from large-scale sequencing. Presently there are over 50,000 STS sequences in EMBL/GenBank [2].

1.1.5. Complete genomic sequences

Since the first completely known genome of *Hemophilus influenzae* was published in 1995 [6], the number of genomes has steadily increased. The hitherto most complex genome is the yeast genome, completed in 1996 [7]. At the end of 1999, 23 genomes are known, and a further 20–30 are expected to be completed within the next 1–2 years. Updated information about ongoing microbial genome projects can be found at the WWW page of TIGR (The Institute for Genomic Research; http://www.tigr.org).

1.1.6. Amino acid sequence databases

For amino acid sequences, the databases differ in their content. Swissprot [8] is a well annotated database with several cross-links to other databases, including those of 2D-PAGE [9], making this database extremely valuable in finding information about proteins. However, since annotation takes time, Swissprot could lag in time compared with those databases with less detailed information. PIR [10] is another protein database with annotations and cross-references which is a continuation of the Atlas project [11]. In contrast to Swissprot, PIR also contains unannotated entries.

Genpept is a protein database with translations of all open reading frames in GenBank, which makes it numerically superior among the protein databases. Similarly, there is a translation of those open reading frames in the EMBL nucleotide database that are not represented in Swissprot. This database is called TrEMBL for translated EMBL [8]. Thus, a search with Swissprot and TrEMBL should be equivalent to a search in Genpept. However, since the database content is not exactly overlapping, it is often wise to search both database sets, especially as the search programs today are fast, and the search is easily performed on a personal computer or via the Internet.

Recent additions and changes to the databases subsequent to the latest full release are available as the smaller databases Swissnew (to Swissprot), TrEMBLnew (to TrEMBL) and GPCU (to Genpept).

In order to make full protein database searches easier, NCBI offers a "nonredundant" database with entries from all amino acid sequence databases and the Brookhaven database of three-dimensional structures (see below). However, this database is not purely nonredundant since partial sequences are treated as separate entries. Thus, searches against this database can result in multiple findings reflecting the same sequence.

Recently, the KIND database has been created [11a], which is truly nonredundant from the sequence point of view.

1.1.7. Databases for three-dimensional structures

Three-dimensional structures of biological macromolecules are collected in the PDB (Protein Data Bank) maintained by the Research Collaboratory for Structural Bioinformatics (RCSB) [12]. They have a WWW site at http://www.pdb.bnl.gov. Most of the structures are determined by X-ray crystallography or nuclear magnetic resonance (NMR), but models are also included in the database. The entries are flat files with lists of x, y and z coordinates for all atoms. The molecules could be displayed on a computer using a molecular graphics program, e.g. Rasmol (available at ftp://ftp.dcs. ed.ac.uk/pub/rasmol) or ICMLite (available via http://www.molsoft.com).

1.2. Integrated user interfaces for databases

SRS (Sequence Retrieval System), developed at EBI and Lion Bioscience, is a user-friendly tool that integrates several different databases [13], containing sequences, three-dimensional structures, sequence patterns, bibliographic data, linkage maps, and much more. SRS offers a WWW interface at http://srs.ebi.ac.uk. There are also several mirror sites available all over the world. At the SRS home page, a list of the current mirror sites is available. SRS offers several cross-links between the databases, which makes it easy for the user to move from a protein sequence to the corresponding nucleotide sequence and then to the Medline entries. Examples of the SRS search forms are shown in Figure 2.

Entrez was developed at NCBI with functionality similar to SRS. Since both systems are constantly updated and improved, regular checks of their services are useful in revealing the pros and cons of each system. It is often a matter of taste and practice which system to use.

1.3. Databases of patterns and protein families

In addition to the sequence databases, there also exist a number of databases containing sequence patterns and sequence alignments, of which some

A
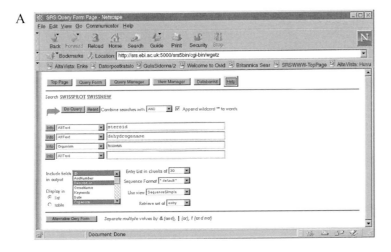

B
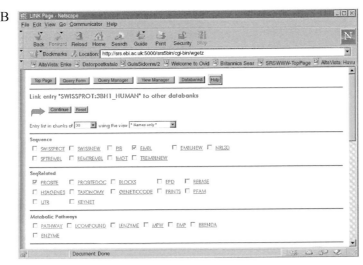

Figure 2. Two SRS search forms. Panel A shows the fill-in form for retrieving data from protein sequence databases. In the example, only human sequence entries containing both a word starting with "steroid" and a word starting with "dehydrogenase" will be retrieved. The result list will display the identification code, description and organism (as selected by the black fields in the "Include fields in output" box). Panel B shows a page for linking an entry (in this example 3BH1_HUMAN) to other databases. The databases of interest are marked by ticking the boxes to the left of the database name (in this example EMBL and PROSITE). Information about each database is retrieved by mouse-clicking on the database name.

common ones are described below. Each of them has its own "philosophy" and preferred usage. Often it is necessary to search in several databases in order to find all the information required. Presently, attempts are ongoing to coordinate these different databases into one to simplify the search tasks.

Prosite is a database containing sequence patterns, characteristic of a protein family or posttranslational modification [14]. The database release of September 1999 contains 1375 patterns corresponding to 1035 families/modifications. These patterns (signatures) are selected to be as specific as possible, and for each pattern the numbers of true positives, false positives, and false negatives are indicated. An example of a Prosite entry is shown in Figure 3. The Prosite database is accessible via http://expasy.hcuge.ch. Available services include tools to scan Prosite for a specific sequence and tools to search the Swissprot protein database for a user-defined pattern. Finally, the Prosite database can also be downloaded for local applications.

Pfam is a database of protein domain families with multiple sequence alignments of their members [15]. The Pfam database is accessible at http://www.sanger.ac.uk/Pfam, from where family entries can be studied and searches made. Pfam contains links to Prosite and, when present, databases of three-dimensional structures.

Blocks are multiple aligned ungapped segments corresponding to the most highly conserved regions of proteins [16]. In this manner, the Blocks database differs from the Pfam database, which contains alignments of complete sequences. The Blocks database is automatically generated from Prosite entries. A WWW service is available at http://www.blocks.fhcrc.org

```
ID   INSULIN; PATTERN.
AC   PS00262;
DT   APR-1990 (CREATED); NOV-1997 (DATA UPDATE); NOV-1997 (INFO UPDATE).
DE   Insulin family signature.
PA   C-C-{P}-x(2)-C-[STDNEKPI]-x(3)-[LIVMFS]-x(3)-C.
NR   /RELEASE=35,69113;
NR   /TOTAL=150(150); /POSITIVE=149(149); /UNKNOWN=0(0); /FALSE_POS=1(1);
NR   /FALSE_NEG=2; /PARTIAL=1;
CC   /TAXO-RANGE=??E??; /MAX-REPEAT=1;
CC   /SITE=1,disulfide; /SITE=2,disulfide; /SITE=4,disulfide;
CC   /SITE=8,disulfide;
DR   P21808, BX4_BOMMO , T; P33718, BXA1_SAMCY, T; P15411, BXA2_BOMMO, T;
DR   P33719, BXA2_SAMCY, T; P26726, BXA3_BOMMO, T; P33720, BXA3_SAMCY, T;
...
DR   P22969, RELX_HORSE, T; P19884, RELX_MACMU, T; Q64171, RELX_MESAU, T;
DR   P01349, RELX_ODOTA, T; P01348, RELX_PIG  , T; P11952, RELX_RAJER, T;
DR   P01347, RELX_RAT  , T; P11953, RELX_SQUAC, T;
DR   P41522, INS_ANGAN , P;
DR   P21563, INS_RODSP , N; P47932, RELX_MOUSE, N;
DR   P50428, ARSA_MOUSE, F;
3D   1BOM; 1BON; IGF2; 1IGL; 1GF1; 2GF1; 3GF1; 1GF1; 2GF1; 3GF1; 2INS; 1APH;
3D   1BPH; 1CPH; 1DPH; 1PID; 1HIQ; 1HIS; 1HIT; 2HIU; 1TYL; 1TYM; 1TRZ; 1HLS;
3D   1HUI; 1BEN; 1AIO; 1AIY; 1LPH; 1XDA; 1XGL; 1MHI; 1MHJ; 1VKS; 1VKT; 3INS;
3D   4INS; 6INS; 7INS; 9INS; 1IZA; 1IZB; 2TCI; 1MPJ; 3MTH; 1DEI; 1SDB; 1WAV;
3D   6RLX; 1RLX; 2RLX; 3RLX; 4RLX;
DO   PDOC00235;
```

Figure 3. The Prosite entry for insulin (panel A) and the corresponding documentation entry (panel B).

{PDOC00235}
{PS00262; INSULIN}
{BEGIN}
****
* Insulin family signature *
****

The insulin family of proteins [1,E1] groups a number of active peptides which
are evolutionary related. This family currently consists of:

 - Insulin.
 - Relaxin.
 - Insulin-like growth factors I and II (IGFs or somatomedins) [2].
 - Mammalian Leydig cell-specific insulin-like peptide (Ley-I-L) (gene INSL3)
   [3].
 - Mammalian early placenta insulin-likepeptide (ELIP) (gene INSL4) [4].
 - Insect prothoracicotropic hormone (PTTH) (bombyxin) [5].
 - Locust insulin-related peptide (LIRP) [6].
 - Molluscan insulin-related peptides 1 to 5 (MIP) [7].

Structurally, all these peptides consist of two polypeptide chains (A and B)
linked by two disulfide bonds.

```
      B chain        xxxxxxCxxxxxxxxxxxxCxxxxxxxxx
                           |               |
      A chain        xxxxxCCxxxCxxxxxxxxCx
                     ***************
                         |   |
                         +----+
```

'C': conserved cysteine involved in a disulfide bond.
'*': position of the pattern.

As shown in the schematic representation above, they all share a conserved
arrangement of four cysteines in their A chain. The first of these cysteines
is linked by a disulfide bond to the third one and the second and fourth
cysteines are linked by interchain disulfide bonds to cysteines in the B
chain. As a pattern for this family of proteins, we have used the region
which includes the four conserved cysteines in the A chain.

-Consensus pattern: C-C-{P}-x(2)-C-[STDNEKPI]-x(3)-[LIVMFS]-x(3)-C
                    [The four C's are involved in disulfide bonds]
-Sequences known to belong to this class detected by the pattern: ALL, except
 for what was thought to be a sponge insulin [8], but which originates from an
 unidentified rodent and which contains sequencing errors and lacks the first
 cysteine of the A chain.
-Other sequence(s) detected in SWISS-PROT: 1.
-Last update: November 1997 / Pattern and text revised.

[1] Blundell T.L., Humbel R.E.
     Nature 287:781-787(1980).
[2] Humbel R.E.
     Eur. J. Biochem. 190:445-462(1990).
[3] Adham I.M., Burkhardt E., Benahmed M., Engel W.
     J. Biol. Chem. 268:26668-26672(1993).
[4] Chassin D., Laurent A., Janneau J.-L., Berger R., Bellet D.
     Genomics 29:465-470(1995).
[5] Nagasawa H., Kataoka H, Isogai A., Tamura S., Suzuki A., Mizoguchi A.,
     Fujiwara Y., Suzuki A., Takahashi S.Y., Ishizaki H.
     Proc. Natl. Acad. Sci. U.S.A. 83:5480-5483(1986).
[6] Lagueux M., Lwoff L., Meister M., Goltzene F., Hoffmann J.A.
     Eur. J. Biochem. 187:249-254(1990).
[7] Smit A.B., Geraerts W.P.M., Meester I., van Heerikhuizen H., Joosse J.
     Eur. J. Biochem. 199:699-703(1991).
[8] Robitzki A., Schroder H.C., Ugarkovic D., Pfeifer K., Uhlenbruck G.,
     Muller W.E.G.
     EMBO J. 8:2905-2909(1989).

1.4. Future

The described databases are not proper databases from a programmer's point of view, but flat files, i.e. normal text files. This has the advantage that the files are easy to read in any program. However, the drawbacks are that the files become larger than otherwise would be necessary and the search possibilities cannot be fully optimised. In the future, we will most likely see an increase of "proper" databases. However, this implies that all search programs need to be simultaneously changed in order to accommodate the new database formats.

Another future direction is the development of systems that make it possible for the data to reside at different physical locations, i.e. distributed databases. This will decrease the Internet traffic for updating databases and simplify the maintenance of each database. However, distribution of complete databases might still be necessary for a limited number of users demanding fast access to the entire database, e.g. for complete genomic and proteomic analysis. Still, important factors for this evolution will be the speed and security of the Internet connections.

Further integration of different databases similar to the initiatives taken by the developers of SRS and Entrez would be helpful for researchers wishing to easily retrieve all available information related to a particular protein. This integration will also successively include more and more databases with different types of information. Furthermore, an integration between sequence databases eliminating the present redundancy would also be valuable to simplify the search procedures.

2. Sequence comparisons

2.1. Comparisons of single sequences towards the databases

When the sequence of a protein or a geneis determined, one immediate task is to search the databases to find whether the sequence is known of whether any related sequence exists from which structural and functional conclusions can be drawn. Today, this can easily be done with locally installed search programs or using the WWW servers at EBI or NCBI. However, sending sequences over the Internet is not secure, and unauthorised people might obtain access to the information. Thus many research laboratories set up their own internal (Intranet) servides. The methods described below are available both as WWW services and as programs to download for local installation.

The FASTA program [17] is based upon modified Needleman-Wunsch [18] and Smith-Waterman [19] algorithms. Two sequences are compared at a time. The program counts both identities and similarities. For the evaluation of similarities, scoring matrices are used. A scoring matrix is an

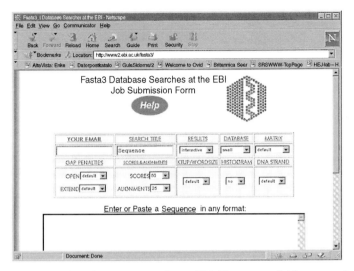

Figure 4. The WWW form for FASTA searches at EBI. First, empty fields are completed and options selected (available alternatives are listed after a mouse click on the down arrow). If the search is not interactive, the e-mail address field must be completed. Second, the sequence must be entered in the large box. The form ignores nonsequence characters and translates lowercase letters to uppercase. Finally, the search is started by a mouse click on the "Run Fasta3" button at the very bottom of the form.

exchange table with all possible amino acid pairs and a value representing the likelihood for that particular exchange, based upon studies of known proteins. Several exchange matrices are used, of which the Dayhoff's PAM250 and Henikoff's BLOSUM62 are among the most frequently used. The FASTA program tries to find the region with highest score between two proteins. Subsequently, this region is extended both N- and C-terminally as long as possible. For insertion of gaps, a penalty is added, typically consisting of one term for the initiation of a gap and one term depending on the length of the gap.

Upon execution of the FASTA program, the user can select several parameters, as can be seen from the WWW form in Figure 4. The most important ones are which database to search, the sensitivity (ktup), gap penalties, and the number of alignments to show. The lower the ktup value, the higher the sensitivity and, consequently, the calculation time. For protein sequences, values of 1 or 2 are recommended, whereas for nucleotide sequences the recommended values are between 3 and 6. The FASTA program makes a global search and reports only the best alignment segment for each protein. Thus, if a protein contains duplicated segments, only one possibility will be shown. The program LFASTA could be used instead to find multiple local alignments.

BLAST [20] is a another set of fast programs for protein and nucleotide sequence comparisons, of which some also allow simultaneous translation

Table 3. Different BLAST programs for different search possibilities

Program	Query sequence	Database
blastp	protein	protein
blastn	nucleotide	nucleotide
blastx	nucleotide, translated × 6	protein
tblastn	protein	nucleotide, translated × 6
tblastx	nucleotide, translated × 6	nucleotide, translated × 6

The indication "translated × 6" means that the sequence is translated in all six possible reading frames before the sequence comparisons.

of nucleotide sequences in all six reading frames (Tab. 3). A WWW service is available at http://www.ncbi.nlm.nih.gov/BLAST with a user interface similar to that described for FASTA above. Recently, PSI-BLAST was created [20] with the possibility to have different score matrices for different positions, which increases the sensitivity when searching distantly related sequences.

2.2. Multiple sequence alignments

The technique to align multiple related sequences is valuable in locating conserved positions. These are often active site residues or other functionally or structurally important residues. These conserved residues can subsequently be studied with site-directed mutagenesis and other experimental techniques.

Most multiple sequence alignment programs are based upon the principle of initial clustering, i.e. to start with pairwise alignment of the sequences most closely related, followed by those second most closely related, and so on. To take all sequences simultaneously into account would be computationally too expensive. However, single programs exist that perform a simultaneous "all-against-all" alignment for a few sequences.

The Clustal programs (presently ClustalW [21] and ClustalX [22]) offer a fast, high-quality alignment procedure combined with a user-friendly menu-driven interface. With Clustal you can calculate multiple alignments and phylogenetic trees (see below). There is also an option to align new sequences with a previous alignment (profile alignment), which is valuable in cases where a large multiple alignment has been arrived at after thorough checking. The program also has possibilities for advanced settings for the experienced user. The ClustalW and ClustalX programs are available for PC, Macintosh, and Unix.

2.3. Phylogenetic tree calculations

Several programs exist for making phylogenetic trees that display relationships between sequences, of which two are exemplified below. To calculate all possible tree topologies to find the one that fits best with the sequence data is still too computationally demanding. Thus the present methods strive at good approximations.

One technique is to calculate the tree that fits best with all pairwise alignment scores. This is used in ClustalW [21]. In order to evaluate the significance of the tree, the program offers the possibility to make bootstrap analyse, i.e. to calculate a large number of trees (e.g. 1000) of which each only considers part of the available data. Subsequently, the branching of these trees is analysed, and the number of trees with branching similar to the original tree is reported. If this number is at least 95% of the number of trees in the bootstrap analysis, the corresponding branching can be regarded as significant at the 95% level.

An alternative technique is to deduce the most parsimonious tree, i.e. the tree topology with the lowest total number of amino acid or nucleotide exchanges. This procedure is used in the program Protpars within the Phylip package [23].

Several tools exist for displaying the evolutionary trees. The Phylip package contains the program Drawtree [23]. The programs njplot [24] and phylo_win [25] for PC/Mac and Unix, respectively, make it possible to interactively change the display order of branches (which does not change the topology of the tree). Further programs for phylogenetic analysis can be found in the software catalogue at EBI (http://www.ebi.ac.uk/biocat/).

3. Prediction of structural properties

3.1. Secondary structure predictions

The Chou-Fasman method [26] uses a combination of rules and statistics (calculated on 4–6 residues) to predict the secondary structure elements α, β, and turn. The Garnier-Osguthorpe-Robson method [27] works similarly but counts statistics on more residues (up to eight on each side of the residue to predict). Both methods have a prediction accuracy of up to 60–65% and are well spread and included in the major sequence analysis packages, e.g. GCG (Wisconsin Package, Genetics Computer Group, Madison, WI, USA). By combining different methods, the success rate could be marginally improved.

The prediction accuracy can be increased by utilising a multiple sequence alignment of related proteins with the same function reflecting the natural and allowed variation [28, 29]. With this technique, the success rate is close to that for neural network methods (e.g. PHD, described below). It is probably difficult to get much beyond that level with this type of method, which only considers the local environment, since the three-dimensional structure depends also on nonlocal interactions.

The PREDATOR method uses the information available from multiple sequences [30]. The program makes pairwise alignments between the query sequence and all other homologous sequences. Only significant alignment fragments are considered in the calculations to avoid prediction errors due to incorrect alignment. The authors report 75% prediction accuracy [30]. There is a WWW-based prediction service available via http://www.embl-heidelberg.de/argos/predator/predator_info.html

PHD is a prediction method using neural networks, a computational technique that attempts to mimic the neuronal connections of the human brain. This technique is powerful in detecting patterns, and several research groups have also applied it to the field of secondary structure prediction. Since the neural network needs to be trained on known sequences and the calculations are computationally intense, these programs are not easily distributed but are executed on one computer, to which users can connect and get their predictions made. The PHD server [31] is widely used and available at http://dodo.cpmc.columbia.edu/predictprotein/. With this technique, a success rate of 72% has been reported [32].

3.2. Membrane protein predictions

Transmembrane segments of proteins are generally hydrophobic. They can thus easily be detected by plotting the hydrophobicity averaged over a number of amino acid residues (usually 6–15) against the position in the primary structure, creating a hydrophobicity plot [33, 34]. Many hydrophobicity scales have been invented, some of which are compared in [35].

Rules for the topology (i.e. which side is intra-/extracellular) of membrane proteins have been studied. One finding is that positive residues have a preference for the inside of membrane proteins – the "positive inside" rule [36, 37]. Multiple sequence alignment techniques have been utilised in order to improve accuracy [38, 39]. Furthermore, neural network techniques have also been applied to this type of prediction [40].

3.3. Prediction of antigenic sites

Several attempts to predict antigenic sites have been made. One of the first approaches was to identify the hydrophilic regions of the protein [33]. In general, the highest hydrophilicity corresponds to one of the antigenic sites. Another approach is to combine several parameters. One method combines hydrophilicity, surface accessibility, flexibility, and turn probability [41].

3.4. Prediction of sites for posttranslational modification

Posttranslational modification serves a number of functions, e.g. protection against proteolysis, membrane attachment, and protein direction after synthesis. Many posttranslational modifications are guided by signals in the primary structure, and in cases where these signals are known, modifications can be predicted. Sequence patterns have been described for acetylation, myristoylation, palmitoylation, phosphorylation, glycosylation, and many more modifications [42–44].

The PROSITE database contains motifs that are easily expressed as a sequence pattern. For patterns based on complicated rules, computer programs for prediction exist that are included in the large sequence analysis packages, e.g. GCG, or could be found on the Internet.

4. Conclusions

Several possibilities exist to deduce information from the sequence alone. Within the near future, we can look forward to the development of further methods and increased usage of all these techniques in the new era of proteomics and bioinformatics.

Acknowledgements
Work in the author's laboratory is supported by the Swedish Medical Research Council and the European Union (BIO4-CT97-2123).

References

1 Stoesser G, Moseley MA, Sleep J, McGowran M, Garciapastor M, Sterk P (1998) The EMBL Nucleotide Sequence Database. *Nucleic Acids Res* 26: 8–15
2 Benson DA, Boguski MS, Lipman DJ, Ostell J, Ouellette BFF (1998) GenBank. *Nucleic Acids Res* 26: 1–7
3 Aaronson JS, Eckman B, Blevins RA, Borkowski JA, Myerson J, Imran S, Elliston KO (1996) Toward the development of a gene index to the human genome: an assessment of the nature of high-throughput EST sequence data. *Genome Res* 6: 829–845
4 Hillier LD, Lennon G, Becker M, Bonaldo MF, Chiapelli B, Chissoe S, Dietrich N, DuBuque T, Favello A, Gish W et al (1996) Generation and analysis of 280,000 human expressed sequence tags. *Genome Res* 6: 807–828
5 Hudson TJ, Stein LD, Gerety SS, Ma J, Castle AB, Silva J, Slonim DK, Baptista R, Kruglyak L, Xu SH et al (1995) An STS-based map of the human genome. *Science* 270: 1945–1954
6 Fleischmann RD, Adams MD, White O, Clayton RA, Kirkness EF, Kerlavage AR, Bult CJ, Tomb JF, Dougherty BA, Merrick JM et al (1995) Whole-genome random sequencing and assembly of *Haemophilus influenzae* Rd. *Science* 269: 496–512
7 Goffeau A, Barrell BG, Bussey H, Davis RW, Dujon B, Feldmann H, Galibert F, Hoheisel JD, Jacq C, Johnston M et al (1996) Life with 6000 genes. *Science* 274: 563–567
8 Bairoch A, Apweiler R (1998) The SWISS-PROT Protein Sequence Data Bank and Its Supplement TrEMBL in 1998. *Nucleic Acids Res* 26: 38–42
9 Hoogland C, Sanchez JC, Tonella L, Bairoch A, Hochstrasser DF, Appel RD (1998) Current Status of the Swiss-2D-PAGE Database. *Nucleic Acids Res* 26: 332–333

10 Barker WC, Garavelli JS, Haft DH, Hunt LT, Marzec CR, Orcutt BC, Srinivasarao GY, Yeh LSL, Ledley RS, Mewes HW et al (1998) The PIR-International Protein Sequence Database. *Nucleic Acids Res* 26: 27–32

11 Dayhoff MO (eds) (1965) *Atlas of protein sequence and structure.* National Biomedical Research Foundation, Silver Spring, Maryland

11a Kallberg Y, Persson B (1999) KIND – a non-redundant protein database. *Bioinformatics* 15: 260–261

12 Abola EE, Bernstein FC, Bryant SH, Koetzle TF, Weng J (1987) *Crystallographic Databases.* Data Commission of the International Union of Crystallography, Bonn/Cambridge/Chester, 107–132

13 Etzold T, Argos P (1993) SRS: an indexing and retrieval tool for flat file data libraries. Computer Applications in the Biosciences 9: 49–57

14 Bairoch A, Bucher P, Hofmann K (1997) The PROSITE database, its status in 1997. *Nucleic Acids Res* 25: 217–221

15 Sonnhammer EL, Eddy SR, Durbin R (1997) Pfam: a comprehensive database of protein domain families based on seed alignments. Proteins 28: 405–420

16 Henikoff S, Henikoff JG (1994) Protein family classification based on searching a database of blocks. *Genomics* 19: 97–107

17 Pearson WR, Lipman DJ (1988) Improved tools for biological sequence comparison. *Proc Natl Acad Sci USA* 85: 2444–2448

18 Needleman SB, Wunsch CD (1970) A general method applicable to the search for similarities in the amino acid sequence of two proteins. *J Mol Biol* 48: 443–453

19 Smith TF, Waterman MS (1981) Identification of common molecular subsequences. *J Mol Biol* 147: 195–197

20 Altschul SF, Madden TL, Schaffer AA, Zhang J, Zhang Z, Miller W, Lipman DJ (1997) Gapped BLAST and PSI-BLAST: a new generation of protein database search programs. *Nucleic Acids Res* 25: 3389–3402

21 Thompson JD, Higgins DG, Gibson TJ (1994) CLUSTAL W: improving the sensitivity of progressive multiple sequence alignment through sequence weighting, position-specific gap penalties and weight matrix choice. *Nucleic Acids Res* 22: 4672–4680

22 Thompson JD, Gibson TJ, Plewniak F, Jeanmougin F, Higgins DG (1997) The CLUSTAL_X windows interface: flexible strategies for multiple sequence alignment aided by quality analysis tools. *Nucleic Acids Res* 25: 4876–4882

23 Felsenstein J (1996) Inferring phylogenies from protein sequences by parsimony, distance, and likelihood methods. *Methods Enzymol* 266: 418–427

24 Perriere G, Gouy M (1996) WWW-query: an on-line retrieval system for biological sequence banks. *Biochemie* 78: 364–369

25 Galtier N, Gouy M, Gautier C (1996) SEAVIEW and PHYLO_WIN: two graphic tools for sequence alignment and molecular phylogeny. *Comput Appl Biosci* 12: 543–548

26 Chou PY, Fasman GD (1978) Prediction of the secondary structure of proteins from their amino acid sequences. *Adv Enzym* 47: 45–148

27 Garnier J, Osguthorpe DJ, Robson B (1978) Analysis of the accuracy and implications of simple methods for predicting the secondary structure of globular proteins. *J Mol Biol* 120: 97–120

28 Persson B, Krook M, Jörnvall H (1991) Characteristics of short-chain alcohol dehydrogenases and related enzymes. *Eur J Biochem* 200: 537–543

29 Levin JM, Pascarella S, Argos P, Garnier J (1993) Quantification of secondary structure prediction improvement using multiple alignments. *Protein Engineering* 6: 849–854

30 Frishman D, Argos P (1997) Seventy-five percent accuracy in protein secondary structure prediction. *Proteins* 27: 329–335

31 Rost B, Sander C, Schneider R (1994) PHD: an automatic mail server for protein secondary structure prediction. *Comput Appl Biosci* 10: 53–60

32 Rost B, Sander C (1995) Progress of 1D protein structure prediction at last. *Proteins* 23: 295–300

33 Hopp TP, Woods KR (1981) Prediction of protein antigenic determinants from amino acid sequences. *Proc Natl Acad Sci USA* 78: 3824–3828

34 Kyte J, Doolittle RF (1982) A simple method for displaying the hydropathic character of a protein. *J Mol Biol* 157: 105–132

35 Degli Esposti M, Crimi M, Venturoli G (1990) A critical evaluation of the hydropathy profile of membrane proteins. *Eur J Biochem* 190: 207–219

36 von Heijne G (1986) The distribution of positively charged residues in bacterial inner membrane proteins correlates with the trans-membrane topology. *EMBO J* 5: 3021–3027

37 von Heijne G (1992) Membrane protein structure prediction. Hydrophobicity analysis and the positive-inside rule. *J Mol Biol* 255: 487–494

38 Persson B, Argos P (1994) Prediction of transmembrane segments in proteins utilising multiple sequence alignments. *J Mol Biol* 237: 182–192

39 Persson B, Argos P (1996) Topology prediction of membrane proteins. *Protein Sci* 5: 363–371

40 Rost B, Casadio R, Fariselli P, Sander C (1995) Transmembrane helices predicted at 95% accuracy. *Protein Sci* 4: 521–533

41 Jameson BA, Wolf H (1988) The antigenic index: a novel algorithm for predicting antigenic determinants. *Comput Appl Biosci* 4: 181–186

42 Persson B, Flinta C, von Heijne G, Jörnvall H (1985) Structures of N-terminally acetylated proteins. *Eur J Biochem* 152: 523–527

43 Eisenhaber F, Persson B, Argos P (1995) Protein structure prediction: recognition of primary, secondary, and tertiary structural features from amino acid sequence. *Critical Reviews in Biochemistry and Molecular Biology* 30: 1–94

44 Han KK, Martinage A (1992) Possible relationship between coding recognition amino acid sequence motif or residue(s) and post-translational chemical modification of proteins. *Int J Biochem* 24: 1349–1363

Subject index